DATE DUE

FE ~~8'01~~			
~~AP 1 8 01~~			

DEMCO 38-296

Life on Other Worlds
The 20th-Century Extraterrestrial Life Debate

The recent discoveries of extrasolar planets and possible microfossils in Martian meteorite ALH 84001 are only the latest developments in a debate that spans millennia and that has been especially heated in the 20th century. From the furor over Percival Lowell's claim of canals on Mars at the beginning of the century to the biological experiments of the Viking spacecraft, the controversial "Mars rock," and the sophisticated Search for Extraterrestrial Intelligence (SETI) at its end, otherworldly life has often titillated and occasionally consumed science and the public. So too have crucially related areas such as the search for planetary systems, the quest for an explanation of UFOs, and inquiries into the origin of life. The theme has been elaborated by science fiction writers from H. G. Wells to Arthur C. Clarke and has resulted in some of the most popular films of all time, including *E.T., Alien, Independence Day,* and *Contact.*

Life on Other Worlds details in a readable and nontechnical manner the history of the 20th-century extraterrestrial life debate, one of the pervasive themes of our century. Unlike other works on the subject, it places the current debate in historical perspective, showing how the concept of extraterrestrial intelligence is a worldview of its own, a "biophysical cosmology" that seeks confirmation no less than physical views of the universe. It is, however, a subject at the very limits of science, and scientific attempts at confirmation therefore illuminate the nature of science itself. This history is not only important for an understanding of the nature of science, but is also central to any forward-looking concept of religion, philosophy, and numerous other areas of human endeavor. Extraterrestrial life will be one of the predominant themes of science in the 21st century.

Steven J. Dick is an astronomer and historian of science at the United States Naval Observatory in Washington, D.C. He is the author of *Plurality of Worlds: The Origins of the Extraterrestrial Life Debate from Democritus to Kant* (Cambridge University Press, 1982), *The Biological Universe: The Twentieth-Century Extraterrestrial Life Debate and the Limits of Science* (Cambridge University Press, 1996), and numerous articles in both scientific and historical journals, including *Space Science Reviews, Journal of the History of Ideas, Technology and Culture,* and *Journal for the History of Astronomy.* Dr. Dick has served as historian for NASA's Search for Extraterrestrial Intelligence (SETI) program and was a member of a NASA workshop examining the cultural aspects of success in SETI, including the short-term and long-term implications of contact with extraterrestrials. He was a member of the panel convened by Vice President Al Gore in 1996 to examine the implications of possible fossilized life in the Mars rock.

"THERE ARE CERTAIN FEATURES IN WHICH THEY ARE LIKELY TO RESEMBLE US, AND AS LIKELY AS NOT THEY WILL BE COVERED WITH FEATHERS OR FUR. IT IS NO LESS REASONABLE TO SUPPOSE, INSTEAD OF A HAND, A GROUP OF TENTACLES OR PROBOSCIS-LIKE ORGANS"

Frontispiece. Illustration by William R. Leigh from H. G. Wells, "The Things That Live on Mars," a nonfiction article that appeared in *Cosmopolitan Magazine* in March 1908 at the height of the Martian canals furor.

Life on Other Worlds

The 20th-Century Extraterrestrial Life Debate

STEVEN J. DICK

CAMBRIDGE
UNIVERSITY PRESS

PUBLISHED BY THE PRESS SYNDICATE OF THE UNIVERSITY OF CAMBRIDGE
The Pitt Building, Trumpington Street, Cambridge CB2 1RP, United Kingdom

CAMBRIDGE UNIVERSITY PRESS
The Edinburgh Building, Cambridge CB2 2RU, UK   http://www.cup.cam.ac.uk
40 West 20th Street, New York, NY 10011-4211, USA   http://www.cup.org
10 Stamford Road, Oakleigh, Melbourne 3166, Australia

© Steven J. Dick 1998

First published 1998

Printed in the United States of America

Typeset in Sabon 10/12 pt., in AMS-TEX [FH]

A catalog record for this book is available from the British Library.

Library of Congress Cataloging-in-Publication Data

Dick, Steven J.
Life on other worlds : the 20th-century extraterrestrial life
debate / Steven J. Dick.
p.   cm.
Includes bibliographical references and index.
ISBN 0-521-62012-0 (hardbound)
1. Mars (Planet) – Exploration.   2. Life on other planets.
3. Exobiology.   I. Title.
QB641.D52   1998
576.8′39 – dc21   98-20465
                                                  CIP

ISBN 0 521 62012 0 hardback

To those who search for the meaning of Life

Glendower: I can call spirits from the vasty deep.

Hotspur: Why, so can I, or so can any man;
           But will they come when you do call for them?

Shakespeare
Henry IV, Part I
Act 3, Scene I, 52–58

CONTENTS

ILLUSTRATIONS AND TABLES

Illustrations

ix

Tables

ACKNOWLEDGMENTS

It is a pleasure to thank once again those who helped with *The Biological Universe,* of which this is an abridgment and update. They include Michael J. Crowe (University of Notre Dame), Ronald Doel, David DeVorkin (National Air and Space Museum), Joshua Lederberg (Rockefeller University), Ronald Schorn, Karl S. Guthke (Harvard University), H. P. Klein (Santa Clara University), Robert Shapiro (New York University), Betty Smocovitis (University of Florida), Philip Klass, David Jacobs (Temple University), Michael Swords (Western Michigan University), and Peter Sturrock (Stanford University).

Among libraries, the unparalleled astronomy collections of the U.S. Naval Observatory Library have been essential for the astronomical portions of this study, as has the assistance of its librarians, Brenda Corbin and Gregory Shelton. In addition, the Library of Congress and the library of The American University have helped fill gaps in nonastronomical literature. I am grateful for access to archives at the British Library, London (A. R. Wallace papers); the Royal Society, London (James Jeans papers); the American Philosophical Society Library, Philadelphia (E. U. Condon and D. H. Menzel papers); Lowell Observatory, Flagstaff, Arizona; University of Arizona, Tucson (A. E. Douglass and G. P. Kuiper papers); Mary Lea Shane archives of the Lick Observatory (Robert Trumpler papers); U.S. Naval Observatory, Washington, D.C. (Clemence papers); and NASA Ames and the SETI Institute in Mountain View, California, for access to SETI archives.

I also wish to thank the SETI Institute for support in undertaking oral history interviews and for the cooperation of all those interviewed. These include John Billingham (NASA Ames), Peter Backus (SETI Institute), David Brocker (NASA Ames), Melvin Calvin (University of California, Berkeley), Gary Coulter (NASA headquarters), Frank Drake (University of California, Santa Cruz), Sam Gulkis (JPL), Nikolai Kardashev, Philip J. Klass, H. P. Klein (Santa Clara University), Michael Klein (JPL), Joshua Lederberg (Rockefeller University), Edward Olsen (JPL), Bernard M. Oliver (NASA Ames), Michael Papagiannis (Boston University), Tom Pierson (SETI Institute), Carl Sagan (Cornell University), Charles Seeger (SETI Institute), Jill Tarter (SETI Institute), and Peter van de Kamp. Oral history interviews of related interest will be found at the Center for the History of Physics of the American Institute of Physics, located at the American Center for Physics in College Park,

Maryland. It is a pleasure to acknowledge the usefulness of David Swift's published interviews in *SETI Pioneers* (University of Arizona Press: Tucson, 1990).

I am grateful to Garland Publishing for permission to draw from my previous article "Plurality of Worlds," *Encyclopedia of Cosmology*, N. Hetherington, ed. (Garland, 1993), 502–512, for Chapter 1 and to Reidel Publishers for permission to use portions of my article "The Search for Extraterrestrial Intelligence and the NASA High Resolution Microwave Survey (HRMS): Historical Perspectives," *Space Science Reviews*, 64 (1993), 93–139, for Chapter 7. I wish to thank those publications allowing me to reproduce illustrations, as stated in the credits, and Suzanne Débarbat (Paris Observatory) and the Juvisy Observatory for help in obtaining the Flammarion photograph. My thanks to Alex Holzman, my editor at Cambridge University Press, and to Helen Wheeler and Helen Greenberg for their help in seeing the volume through the production process.

Finally, thanks once again to my wife, Terry, and my sons, Gregory and Anthony, who continue with me to explore new worlds.

INTRODUCTION

The eternal silence of these infinite spaces frightens me.

Pascal, *Pensées*

Deep in the summer of 1996, a startling announcement came from the National Aeronautics and Space Administration (NASA), the American space agency. Life had been found on Mars! Maybe. The very possibility set the world afire, igniting media hype, public imagination, and scientific curiosity alike.

On August 7, with little more than a day's notice, reporters descended on a hastily called NASA press conference, to which the participants themselves had been hurriedly summoned. A carefully planned announcement for the following week had been upstaged by a three-paragraph leak in the industry newspaper *Space News,* and the exhausted scientists had flown in from around the country. Among the many officials in the audience were the heads of the National Science Foundation (NSF), and the National Academy of Sciences, and Gerald Soffen, the project scientist for the Viking spacecraft, which had landed on Mars 20 years earlier. First to the podium was NASA Administrator Dan Goldin, who had already briefed President Bill Clinton and other top political officials. He waxed eloquent about NASA, American science, and the breathtaking conclusions about to be announced, and reported that the president had asked that the discovery be given top priority and that Vice President Al Gore call a space summit to examine its significance. Indeed, President Clinton professed, "I am determined that the American space program will put its full intellectual power and technological prowess behind the search for further evidence of life on Mars."

NASA Associate Administrator for Space Science Wes Huntress then turned the podium over to the scientists, a team of nine led by geochemist David McKay of NASA's Johnson Space Center. Now they presented their evidence to a hushed audience. Organic molecules had been found in a meteorite that was blown off of Mars 16 million years ago, had landed in the Antarctic 13,000 years ago, was found there by a meteorite-collecting team funded by NSF and the Smithsonian Institution 12 years ago, and had been recognized as Martian only 2 years ago. Two years of exhaustive study had led the researchers to their momentous conclusions. The claim of organic molecules on Mars was already a step beyond the Viking results. But there was much more: mineral "carbonate globules" of possible biological origin; evidence of tiny magnetic minerals that on Earth are secreted by certain bacteria; and finally, pictures

I

of strange elongated, hauntingly wormlike structures that they argued might be microfossils. In short, the assembled scientists suggested, life had existed on Mars sometime in the planet's distant past, when Mars was warmer and wetter.

This was not exactly the Martian civilization some had hoped for, but compared to the dry results of Viking 20 years before, it was little short of miraculous. The result had already been peer-reviewed and was scheduled to appear in the August 16 issue of the prestigious journal *Science*. Yet, as always in the debate over life on Mars, there was a "maybe," and NASA had arranged for skeptical University of California at Los Angeles (UCLA) paleobiologist William Schopf to comment at the news conference. He agreed that the evidence of Martian origin was good, and that the evidence of organic molecules was good though not proved to be extraterrestrial. But he argued that the fossils were 100 times smaller than the smallest such fossils found on Earth and were not proved to be fossils at all. He quoted Carl Sagan, the pioneering exobiologist who died only a few months later, that "extraordinary claims require extraordinary evidence" and concluded that incontrovertible evidence for life (e.g., a cell) had not yet been found. Thus began yet another debate over Martian life, a subject that has exerted a peculiarly romantic pull on popular and scientific imagination for more than a century.

In subsequent months, British scientists confirmed the existence of organic molecules in a much younger Martian meteorite. But the claim of Martian fossils proved to be a much harder sell, especially when critics claimed that the carbonate globules had been formed at temperatures so high that life could not have been associated with them. Nevertheless Sagan and others proclaimed a new era, and most Americans, at least, were ready to follow. Whether or not the Mars rock proved in the end to harbor life, one thing was clear: the rock dramatically drew the attention of millions to the fact that Mars had once been warmer and wetter, the kind of place where life might have developed. Life on Mars, a subject practically dead since the Viking missions of 1976, was given a new lease on life and seemed destined to be one of the major themes of the 21st century. Already in 1997 the Pathfinder spacecraft landed on Mars with its roving Sojourner, and the Mars Global Surveyor arrived to chart a new world. And these were only the vanguard of many more spacecraft to come, always with that tantalizing question of life in mind.

The year 1996 was a turning point in other areas related to the extraterrestrial life debate. Even as the significance of the Martian meteorite was being pondered, the Galileo spacecraft returned high-resolution pictures of the enigmatic Jovian satellite Europa, showing a surface likely to be cracked ice, probably floating on an enormous ocean, which could contain life. Beyond the solar system, 1996 also saw the confirmation of one of the Holy Grails of astronomy of the 20th century: the discovery of not only 1 but 8 (and possibly 10) planets orbiting other Sun-like stars. Although not Earth-like, they fueled further the fires of extraterrestrial expectation. More generally, life was increasingly found flourishing on Earth in extreme environments – notably in hydrothermal vents at temperatures of 110°C, inside rocks, and several kilometers below the Earth's surface, requiring energy sources and metabolisms alien to our usual ways of thinking. If life could survive in such extreme conditions, exobiological optimists asked, could it not thrive in a variety of planetary environments in outer space? It all seemed very logical even to the most jaded skeptic.

All of these events were only the most recent manifestation of a debate that stretches back through millennia of history. My own interest in these discoveries was sharpened because my history of the extraterrestrial life debate during the 20th century had been published only a few months before. *The Biological Universe: The Twentieth-Century Extraterrestrial Life Debate and the Limits of Science* said very little about fossil life on Mars because it had simply not yet played an important part in the debate. While Europan seas had received some treatment among scientists and the press, the idea had not yet received the widespread attention it now commands in science and will command in history. Nor was the outcome of the planetary systems debate at all as clear as it now seems to be. With the conviction that historical perspective is essential to understanding the current debate, I have with this volume abridged *The Biological Universe* in an attempt to reach a wider audience and have also taken the opportunity to integrate the new discoveries into the story. Altogether, *Life on Other Worlds* offers not only an interpretation of the events in the debate, but also an interpretation of their significance, which in my judgment constitutes a world view comparable to the Copernican and Darwinian worldviews.

Meanwhile, the influence of extraterrestrial life on popular culture, already a major theme of the 20th century, shows no sign of flagging. The movie *Independence Day*, in which humanity fights off the invading aliens, was by far the most popular movie of 1996, even as *Men In Black* and the movie version of Carl Sagan's *Contact* were among the most popular movies of 1997. The suicide in California of 39 individuals who believed they were going to beam aboard an alien spaceship following in the wake of comet Hale–Bopp in the spring of 1996 shows the tragic outcome of a cosmotheology gone wrong. The Pathfinder landing on Mars on July 4, 1997, was juxtaposed with media coverage of the thousands who gathered in Roswell, New Mexico, to celebrate the 50th anniversary of the supposed crash of an alien spaceship. And as we approached the new millennium, unidentified flying object (UFO) reports were once again on the rise, and major polls showed that most people believe them to be spaceships piloted by extraterrestrials. It is a long way from Martian microfossils and extrasolar planets to invading aliens, but in this volume the reader will find background for the whole range of issues in the extraterrestrial life debate, up to and including the search for intelligence and concepts of aliens in fact and fiction.

Whether or not one ascribes all this interest to deep psychological yearnings for companionship and superior wisdom, the subject is of interest not only because it is believed by such a large portion of the population, but also because it has been seen as amenable to contemporary science, which has struggled at the limits of its capabilities to resolve the many issues associated with the debate over life beyond the Earth. Nor is the subject limited to science and popular culture; no matter how much we learn about the varied life forms of Earth and the physical nature of the universe of which we are a part, the question of our biological uniqueness in the universe is central to the quest for who we are and what our role in nature may be, questions as much a part of religion and philosophy as of science. As Harvard Professor Karl Guthke has claimed in his book *The Last Frontier,* the question of extraterrestrial life is one of the most important myths of the modern age, where *myth* is used in the broad sense of a symbolic tradition that defines how we understand ourselves. Moreover, the history of the pursuit of extraterrestrials is also a key to the nature of science, a field that dominates our culture. The debate provides an unusual opportunity to examine scientific activity, inference, and community when the subject of study is at the very limits of science.

On the most basic level, then, this book is a study of a persistent theme in 20th-century Western culture, a theme that is reflected in other cultures to an extent that still needs to be determined. More than that, it is the story of the transformation of human thought from the physical world to the biological universe, a story no less profound than the move from the closed world to the infinite universe that historians well documented decades ago. It is, therefore, the saga of one of the major worldviews of the 20th century. That the majority of humans now view themselves as surrounded by celestial intelligences, either on Earth or in outer space, surely defines the modern world no less than did the choirs of angels and legions of devils in the medieval world. And the persistence of these supernatural intelligences in some religions of the modern world leads one to believe by analogy that extraterrestrials will persist in human consciousness, even lacking definitive evidence of their existence.

Those who wish to know the full history of the extraterrestrial life debate prior to the 20th century may consult (in addition to Guthke) my own *Plurality of Worlds: The Origins of the Extraterrestrial Life Debate from Democritus to Kant* (1982) and Professor Michael J. Crowe's *The Extraterrestrial Life Debate 1750–1900: The Idea of a Plurality of Worlds from Kant to Lowell* (1986), both published by Cambridge University Press. And those who wish to find complete documentation for this abridged history will find ample footnotes for all but the most recent events in *The Biological Universe*. Should this abridgment help a wider audience to understand and place in context one of the most deeply rooted themes of our time, it will have served its purpose.

FROM THE PHYSICAL WORLD TO THE BIOLOGICAL UNIVERSE
Democritus to Lowell

Astronomy has had three great revolutions in the past four hundred years: The first was the Copernican revolution that removed the earth from the center of the solar system and placed it 150 million kilometers away from it; the second occurred between 1920 and 1930 when, as a result of the work of H. Shapley and R. J. Trumpler, we realized that the solar system is not at the center of the Milky Way but about 30,000 light years away from it, in a relatively dim spiral arm; the third is occurring now, and, whether we want it or not, we must take part in it. This is the revolution embodied in the question: Are we alone in the universe?

Otto Struve (1962)

From the ancient Greek world of Democritus to the 18th-century European world of Immanuel Kant, cosmological thinking underwent a revolution that transformed a dead celestial world into a living universe, a transformation no less dramatic than the well-known move from the closed geocentric world to the infinite universe. These two revolutions – from the closed world to the infinite universe and from the physical world to the biological universe – are not unrelated. The shattering of the ancient world of nested spheres not only opened the way for a greatly enlarged and interrelated universe by removing the Earth from its central status, it also opened the question of the uniformity of natural law. And the same logic that argued for the uniformity of physical law throughout the universe might also be applied to the uniformity of biological law, producing not only planets, stars, and stellar systems but also plants, animals, and intelligence. But while the first could be proved by appeal to theory and observation, the second remains to be proved to the present day. The biological universe therefore is still a revolution of the mind, a transformation that seeks confirmation in the real world, a task that the 20th century has taken to heart.

How, then, did the transformation from the physical world to the biological universe take place, and how has its central thesis of universal life been propagated into modern consciousness? It is the contention of this chapter that this occurred in three stages over two and a half millennia. The idea of life beyond the Earth was first initiated, and sustained through the 17th century, by a variety of cosmological worldviews. Within the framework of the surviving cosmologies, the debate was then refined and propagated through the middle of the 19th century by both religious and secular philosophies. And finally, in the second half of the 19th century, it achieved its scientific

foundations in two overriding developments – the theory of biological evolution and the rise of astrophysics. At the same time it became increasingly
susceptible to the methods of science, a pattern that leads directly to the 20th-
century debate.

In attempting to characterize such a broad period, we must be careful not
to draw too rigidly the boundaries of science and philosophy. For many centuries they were inextricably intermixed; indeed, what we today call *science*
emerged from *natural philosophy*. In claiming that the debate has moved
from the great generalizations of physical cosmology, to the exploration of
philosophical implications, to more empirical investigations, we must constantly keep in mind that the philosophical is never banished completely and
that the cosmological is always present in the background. This is equivalent to stating what is widely accepted: that the subject of extraterrestrial life
has become more amenable to the methods of modern science – observation,
theory, and experimentation – while still being enmeshed in philosophical assumptions that all of science seems unable completely to escape. Indeed, the
extraterrestrial life debate itself may be seen as a struggle for a worldview,
with all of the problems that this implies. Such a status goes a long way toward explaining the emotional nature of the debate, for much more is at stake
than another scientific theory. Given its scientific components, the idea of a
universe filled with life is in fact a cosmology of its own, incorporating the
physical and the biological, and which we may therefore term the *biophysical
cosmology*.

I.I THE COSMOLOGICAL CONNECTION

The extraterrestrial life debate began as part of cosmological worldviews, and
if we take *cosmological* to have its broadest meaning, this may also include
mythical cosmologies. The idea that life might exist beyond the Earth undoubtedly dawned in human consciousness long before it appeared in clearly
recognizable form in the Greek cosmology of the 5th century B.C. Perhaps in a
rudimentary way it appeared among the gods and goddesses who peopled the
heavens of ancient mythology. Sentient beings beyond the Earth were probably part and parcel of early human attempts to understand nature through
rudimentary theology; in this fact is found another reason for the emotional
attachment to the issue in recorded history to the present day.

Such a view might also help to explain why the idea of other worlds was already present in the earliest scientific cosmological worldview, ancient atomism. Constructed in the 4th and 5th centuries B.C. by Leucippus, Democritus, and Epicurus, atomism was, of course, more than a cosmology. But
the important point that we shall see borne out again and again in these
cosmological worldviews is this: to a large extent the concept of other worlds

was based on the physical principles of the cosmological system. So it was with ancient atomism, in which Epicurus spoke of an infinite number of worlds resulting from an infinite number of atoms. The "world" of Epicurus was the Greek *kosmos,* meaning an ordered system, as opposed to chaos. The entire visible universe composed one kosmos; Epicurus here proposed the remarkable idea that an infinite number of such worlds exist completely beyond the human senses – but not beyond human reasoning. For according to the atomist system, the infinite number of atoms could not have been used in our finite world. As our world was created by the chance collision of atoms in an entirely natural process, so must other worlds be created in like manner.

This atomist doctrine would eventually be spread throughout Europe by the Roman poet Lucretius (ca. 99–55 B.C.), whose *De rerum natura* (On the Nature of Things) supported the belief of Epicurus: other worlds must exist "Since there is illimitable space in every direction, and since seeds innumerable in number and unfathomable in sum are flying about in many ways driven in everlasting movement." Moreover, Lucretius added that nothing in the world was unique, including the world itself, and "when abundant matter is ready, when space is to hand, and no thing and no cause hinders, things must assuredly be done and completed." It is notable here, however, that what can be done is defined by the physical principles that one accepts. It is almost inconceivable that the concept of infinite worlds would have arisen in the atomist system in the absence of the physical principles of the atomist cosmology.

Those physical principles, however, were not destined to win the day or even the millennium; it was almost 2000 years before they would be revived in the 16th and 17th centuries with the birth of modern science. In the meantime, a far more elaborate cosmology was constructed by Aristotle (383–322 B.C.), student of Plato and founder of the Lyceum in Athens, whose life overlapped that of Epicurus by two decades and who gave new meaning to the word kosmos. Aristotle's cosmology placed the Earth at the center of a nested hierarchy of celestial spheres, from the spheres of the Moon and planets to the sphere of the fixed stars. The Earth was ever-changing and corruptible, as could be seen by experience, while the region of the celestial crystalline spheres was eternally unchanging. And the Earth was more than a physical center; it was also the center of motion. According to one of the basic tenets of Aristotle's cosmology – the doctrine of natural motion and place – everything in the cosmos moved with respect to that single center: the element earth moved naturally toward the Earth, and the element fire moved naturally away, while air and water assumed intermediate natural places. Aristotle's belief in the impossibility of more than a single kosmos was directly tied to this basic tenet. In his cosmological treatise *De caelo* (On the Heavens) he reasoned that if there were more than one world, the elements of earth and fire would have more than one natural place toward which to move, a physical and logical contradiction.

The issue of a plurality of worlds was thus reduced to a confrontation with the most basic assumptions of Aristotle's system. Either he must reject his doctrine of natural motion and place, on which he had built his entire physics, along with his belief in four elements, on which his theory of matter rested, or he must conclude that the world was unique. The choice was not difficult; indeed, he must have taken comfort in reaching a conclusion so diametrically opposed to that of the atomists, whose system differed from his in so many other ways.

It was Aristotle's system that was transmitted to the Latin West, where it was repeatedly commented on in the context of the Christian system. For Christianity a plurality of worlds directly confronted its omnipotent God: suppose God wished to create another world; how could he do so given the principles of Aristotle? Either Aristotle was wrong, and by his own admission wrong in some very basic principles, or God's power was severely limited. For the first century of Aristotle's introduction into the West, his conclusion of a single world was largely accepted and the concept of God's omnipotence redefined. Thomas Aquinas, among others, argued that God's perfection and omnipotence could also be found in the unity of the world. But by the 13th and 14th centuries, university scholars claimed that the plurality of worlds was not theologically impossible because God can act beyond the Aristotelian laws of Nature. God could create another world and reorder its elements so that they would move according to Aristotelian law but with respect to their own world. William of Ockham – of Ockham's razor fame – went still further by altering Aristotle's doctrine of natural place to state that the elements in each world would return to the natural place within their own world, without any intervention from God. By 1377 the Paris master and bishop of Lisieux, Nicole Oresme, had completely reformulated the doctrine of natural place to state that as long as heavy bodies were situated in the middle of light bodies, no violence would be done to the doctrine of natural place. In one stroke, he thus abandoned the Earth–outer sphere relation so central to Aristotle. Other worlds were possible, and without any supernatural intervention. But while possible, almost all the medieval Scholastics stressed, God in reality had not created more than one world. Neither the modern doctrine of the plurality of worlds nor the universe of modern science emerged from the Middle Ages, despite significant advances over Aristotle.

Like the move to an infinite universe, the transition to the rudiments of a biological universe was not made through successive rebuttals to Aristotle's doctrine of a single world. Rather, it stemmed from the complete overthrow of Aristotle's geocentric universe and its replacement with the Copernican system of the world. It is true that Copernicus did not discard completely the old worldview, especially its epicyclic explanations for certain planetary motions. Nor was the Copernican worldview a general cosmology that was meant to explain the entire universe. But by placing the Sun in the center of the system of the planets and making the Earth one of those planets, he laid the foundations for future cosmologies that would be more general and would incorporate the heliocentric system.

It was this heliocentric system that gave birth to the new tradition of the plurality of worlds, where *world* (*mundus*) was redefined to be an Earth-like planet, which now took on the kinematic, or motion-related, functions of the single Earth in the old geocentric system. Just as the kinematic implications of the decentralization of the Earth led to the birth of a new physics, so the "planetary physics" implications of that move led to the birth of the concept of the biological universe. All discussions of life on other worlds since then, whether consciously or not, recall that fainter echo that Copernicus set in motion: if the Earth is a planet, then the planets may be Earths; if the Earth is not central, then neither is humanity.

Copernicus himself did not pursue the implications of his system, but the ideas of the Italian philosopher Giordano Bruno showed just how far such implications might go, though he was largely influenced by more metaphysical arguments. An avowed Copernican, Bruno nevertheless gave himself full credit for his universe filled with inhabited worlds. In his *De l'infinito universo e mondi* (On the Infinite Universe and Worlds, 1584), Bruno pointed to the metaphysical concept of unity as the source of his belief. The unity of the universe shattered the old Aristotelian spheres, as well as any celestial–terrestrial dichotomy, and it led to innumerable worlds via his conviction that both the greatness of Divine power and the perfection of Nature lay in the existence of infinite individuals, including infinite worlds. Greatly influenced by the atomists, their principle of plenitude also entered the argument: Nature could not help but produce infinite worlds. For his belief in other worlds, as well as other supposed heresies, the Roman Catholic Church burned Bruno at the stake in 1600 (Fig. 1.1).

Though not tied directly to his Copernicanism, Bruno's was a view that would undoubtedly have profoundly disturbed Copernicus. And yet, it was the view toward which his system inexorably led in a march that had not been completed by the 20th century. Even before the invention of the telescope, which would reveal the roughly Earth-like nature of the Moon and at least some of the planets, the young astronomer Johannes Kepler, already a convinced Copernican, would ascribe inhabitants to the Moon. The invention of the telescope began the long trend of attempts at empirical verification of the Copernican implication that the planets were Earth-like. The question was, how much like the Earth? In the first announcement of his observations in his

Fig. 1.1. The 51-year-old Giordano Bruno being burned at the stake on February 17, 1600, by order of the Inquisition. The scene is depicted at the base of the statue of Bruno in the Piazza Campo dei Fiori (Field of Flowers) in Rome, the site of Bruno's burning. It is now surrounded by a colorful and festive market.

Siderius nuncius (Sidereal Messenger, 1610), Galileo noted that the surface of the Moon was "not unlike the face of the earth," and that its bright parts might represent land and its darker parts water. But within a matter of weeks Kepler had responded to Galileo with his own publication. Not only did he agree with Galileo's interpretation of the dark and bright lunar spots, he also had a remarkable explanation for a large and particularly circular lunar cavity detected by Galileo: it was formed by intelligent inhabitants who "make their homes in numerous caves hewn out of that circular embankment." Kepler also argued for a lunar atmosphere, and in his *Somnium* (*Dream, 1634*) would later detail the nature of the lunar inhabitants. Beyond the moon Kepler could only resort to more philosophical arguments, aside from the general Copernican implication. The newly discovered moons of Jupiter, for example, could not be meant for the inhabitants of Earth, who never see them, but for the Jovians.

The response of Aristotelians to such ideas was outrageous astonishment, and even among Copernicans predictable caution was the watchword. Galileo himself at first denied such implications and even in his defense of the Copernican system (*Dialogue on the Two Chief World Systems, 1632*) sought to downplay the similarities of Earth and Moon, admitting only that if there was lunar life, it would be "extremely diverse and far beyond all our imaginings." The Copernican tide, however, could not be stemmed. Six years later, in the less repressive atmosphere of Anglican England, Bishop John Wilkins penned his *Discovery of a World in the Moone* (1638), where Galilean caution was thrown to the wind, while avoiding some of the wilder claims of the irrepressible Kepler. Making use of the few observations available, as well as theology and teleology, Wilkins argued strongly for an atmosphere around the Moon and agreed with Galileo that the dark spots were water. And he did not fail to make clear the inspiration for his work; its tenets, he held, could be deduced from Copernicus and his followers, "all who affirmed our Earth to be one of the planets, and the Sun to be the center of all, about which the heavenly bodies did move. And how horrid soever this may seem at first, yet it is likely enough to be true." Copernicanism was not synonymous with inhabited planets, but it did give theoretical underpinning to habitable planets. The proof or disproof of this implication remained a goal of astronomers until the Viking landers touched down on Mars in the late 20th century.

All important cosmological worldviews of the 17th century and later incorporated the Copernican system as a basic truth. Such was the case with the first complete physical system proposed since Aristotle, that of the French philosopher René Descartes. His *Principia Philosophiae* (Philosophical Principles, 1644), also greatly influenced by a revived atomism, offered a mechanical philosophy in which atoms in motion once again formed the basis for a rational cosmology. For the plurality of worlds tradition it did even more, for it was through the Cartesian cosmology that the quest for a biological universe was first carried to other solar systems, and in a fashion so graphic that it remained an ingrained concept to the present day. Unlike the void space of his atomist predecessors (and his Newtonian successors), Descartes proposed that the universe was a plenum, filled with atoms in every nook and cranny. A consequence of this was that, once set in motion by God, the particles of the plenum formed into vortices, systems analogous to our solar system, centered around every star. Though Descartes himself, again for religious reasons, was careful not to specify that these vortices consisted of inhabited planets, his application of Cartesian laws to the entire universe and the graphic vortex cosmology was plain for all to see.

Descartes's followers were not slow to realize the implications, and some of them elaborated on the nature of the whirling vortices. But none was more bold or more successful than his countryman Bernard le Bovier de Fontenelle. His *Entretriens sur la pluralité des mondes* (Conversations on the Plurality of Worlds, 1686), a treatise that exploits both the Copernican and the Cartesian theories to shed light on the question of life on other worlds, was explicit in its reason for spreading life beyond the solar system:

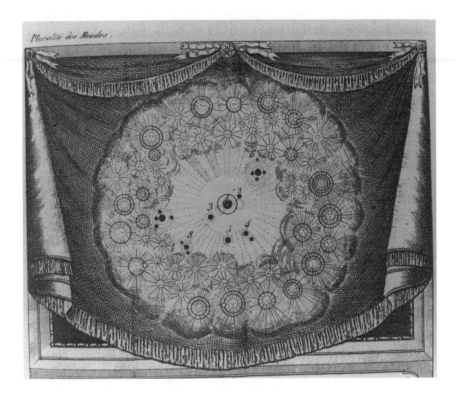

Fig. 1.2. Frontispiece to the 1821 French edition of Fontenelle's *Entretriens sur la plu-ralité des mondes* (1686), depicting the plurality of solar systems. From Steven J. Dick, *Plurality of Worlds* (Cambridge, 1982), by permission of Cambridge University Press.

> If the fix'd Stars are so many Suns, and our Sun the centre of a Vortex that turns round him, why may not every fix'd Star be the centre of a Vortex that turns round the fix'd Star? Our Sun en-lightens the Planets; why may not every fix'd Star have Planets to which they give light?

The *Entretriens* spread throughout Europe in an extraordinary number of editions, and with it the idea of a plurality of solar systems (Fig. 1.2) became ingrained in the European consciousness. In the very same year the famous Dutch astronomer Christiaan Huygens began to formulate his own ideas on the plurality of worlds, published posthumously in the *Kosmotheoros, sive, de terris coelestibus earumque ornatu conjecturae* (Cosmotheoros, or, Conjec-tures concerning the Celestial Earths and their Adornments, 1698). Strongly motivated by his experience as an observational astronomer and explicitly building on the foundation of the Copernican theory, Huygens also proposed a plurality of solar systems, based primarily on the analogy that the fixed stars were suns (another of Descartes's tenets) and only loosely tied to Cartesian vortices. Although Cartesian vortices would soon be swept away by the Newtonian system, the general idea of planetary systems would not.

It is ironic that, of all these cosmological worldviews, the scientific principles of the Newtonian worldview – the system that we have inherited in modified form – entailed extraterrestrial life least of all. The ancient atomists held that an infinite number of atoms in an infinite universe must necessarily form an infinite number of worlds, given the example of our finite world; the Copernican principle of the noncentrality of the Earth led directly to the implication of other Earths; and the Cartesian plenum led directly to vortices that common sense dictated were similar to our own solar system. Although a mechanical philosophy like that of Descartes, Newton's atoms and void, with each body subject to universal gravitation according to fixed laws, did not necessarily imply other solar systems. There was no mechanical necessity for the formation of systems, as there had been in Descartes's system; indeed, under Newtonian principles the whole question has proved to be one of the greatest complexity to the present day.

One could, of course, argue that since our solar system exists and the laws of gravitation are universal, other solar systems should exist. This, however, was only the grossest analogy, almost equivalent to assuming what one wished to prove. Newton himself declined to expound any rational cosmogony that might shed light on the question; in fact, in his letters to the theologian Richard Bentley he indicated that he did not consider it possible to settle this question based on the principles of his system. Instead, Newton insisted that the formation of all ordered systems was contingent on God's will, contenting himself with the observation in the second edition of his Principia (1713) that "if the fixed stars are the centres of other like systems, these, being formed by the like wise counsel, must be all subject to the dominion of the One. . . ."

Despite the arguments of Newton and others to retain the Deity in a system subject to natural law, it became increasingly clear that his laws actually lessened the need for a Deity. In this atmosphere it is not surprising that the concept of a plurality of worlds was put to good use – as a proof of God's glory drawn from his works in Nature, the enterprise known as *natural theology*. If God could no longer be given a role in maintaining his universe, then the concept of a plurality of inhabited planets could be made to reflect the glory, power, and wisdom of the Creator. In one Newtonian treatise after another, the theological view of an inhabited universe was joined to the physical principles of Newton's system. Again and again, a universe full of inhabited solar systems was applauded as one far more worthy of the infinite Creator than any of the other narrower schemes. Once this decision had been made, overwhelming all Scriptural objections, other arguments could be adduced in its favor. One of the most frequent was the argument of teleology, of purpose in the universe, clearly set forth already by Bentley when he wrote "All Bodies were formed for the Sake of Intelligent Minds: As the Earth was principally designed for the Being and Service and Contemplation of Men; why may not all other Planets be created for the like uses, each for their own Inhabitants who have Life and Understanding."

This satisfying vision of the universe, operating by Newtonian laws and reflecting the power of the Deity by spreading intelligence through the universe, was transmitted to the modern world. The proof of other solar systems by observation, and the proof of their likely formation by Newtonian principles, remained a desired goal in the centuries to follow. But the basic predisposition toward a universe of inhabited solar systems was set, almost within the lifetime of Newton himself. Indeed, the Newtonian system was carried to its ultimate extreme in this respect already in the 1750s and 1760s when the philosopher-cosmologists Thomas Wright, Immanuel Kant, and Johann Lambert argued for a hierarchy of ordered systems stretching from our own solar system to the system of the Milky Way Galaxy itself and even beyond. Their universe, "animated with worlds without number and without end," brings us back almost full circle to the ancient Greek atomist view of infinite kosmoi, systems of stars separated from one another, each filled with a multitude of inhabited worlds. But now the laws of motion of such systems had been fathomed, somehow extracted from Nature by the very mind that was now projected throughout the universe.

1.2 PHILOSOPHICAL EXPLORATIONS

Following the triumph of the Newtonian system in the mid-18th century, the extraterrestrial life debate was waged not so much on a cosmological scale as on a scale of worldviews a level or more below the cosmological. Though sometimes discussed by the elaboration of Newtonian science such as the Laplacian nebular hypothesis, more often it fell to the domain of philosophical explorations, both secular and religious. If cosmological worldviews gave birth to the idea of extraterrestrial life, then philosophy and literature, in their traditional role of examining the human condition, explored the ramifications of the idea borne of that cosmological context.

In particular, much of the late-18th-century and 19th-century plurality-of-worlds debate – at least in the West – may be understood as a struggle with that widespread philosophical worldview known as Christianity. If in the Newtonian system the plurality of worlds concept was reconciled with theism via natural theology, this was not equivalent to a reconciliation with

Christianity; as Professor Crowe succinctly states in *The Extraterrestrial Life Debate, 1750–1900,* "structures of insects or solar systems may evidence God's existence, but they are mute as to a Messiah." Three choices were logically open to Christians who pondered the question of other worlds: they could reject other worlds, reject Christianity, or attempt to reconcile the two. Historically, all three of these possibilities came to pass in the 18th and 19th centuries.

Though the Scriptural and doctrinal problems of the issue had been widely discussed throughout the 17th century, only to be overwhelmed by natural theology, no one expressed the continuing difficulties of the plurality of worlds doctrine for Christianity more forcefully than Thomas Paine, that agitator for freedom on two continents. In his influential *Age of Reason* (1793), which saw numerous American editions by the end of the century, Paine bluntly stated that "to believe that God created a plurality of worlds at least as numerous as what we call stars, renders the Christian system of faith at once little and ridiculous and scatters it in the mind like feathers in the air. The two beliefs cannot be held together in the same mind; and he who thinks that he believes in both has thought but little of either." With millions of worlds under his care, Paine argued, could we really believe that the Messiah came to save human beings on this small world? Or did the redeemer hop from one world to the next, when the number of worlds was so great that he would be forced to suffer "an endless succession of death, with scarcely a momentary interval of life"? Despite the force of this argument, few would reject Christianity as did Paine. But few would reject plurality of worlds either, a testimony to its entrenchment by the end of the 18th century. This left but one alternative: the two systems and all they implied would have to coexist.

That other worlds could be incorporated into Christianity, despite Paine, was demonstrated by the Scottish theologian Thomas Chalmers. His *Astronomical Discourses* (1817) incorporated plurality of worlds into evangelical religion, and his countryman Thomas Dick made it a staple of Christianity in a number of works during the first half of the 19th century. But Paine's objections would not disappear. By midcentury the compatibility of plurality of worlds with Christianity was once again called into serious question, this time by William Whewell, philosopher, scientist, and Master of Trinity College, Cambridge. His treatise *Of a Plurality of Worlds: an Essay,* which appeared anonymously in 1853, was the most learned, radical, and influential antipluralist treatise of the century. Unlike Paine and others, Whewell argued that it was pluralism, not Christianity, that should be rejected. To the argument that all the vast space must have some purpose, he countered that geology reveals human existence on Earth to be but a short "atom of time" compared to the age of the Earth; therefore, why could not intelligence be confined to the "atom of space" that was the Earth? And although the universe was indeed vast (about 3000 light years by his estimate), Whewell argued that the possible locales for inhabitants had been vastly overrated. The stars might not be like the Sun, the planets might not be like the Earth; in short, Whewell saw the analogies as greatly exaggerated in the case of other worlds. No longer was the Copernican fact that the planets were Earths a sufficiently precise argument; greater attention had to be given to the details of their physical conditions. Whewell's treatise generated a tremendous amount of debate, but in the end it did little to weaken support for a plurality of worlds among scientists or the religious.

Innumerable other discussions of the relation of plurality of worlds to Christianity were penned throughout the 19th century. Almost without fail, the point of all of them is some kind of reconciliation, with an occasional conclusion that one or the other would have to go. Reconciliation with the doctrines of Incarnation and Redemption was never achieved, however, with most insisting that Christ's incarnation on Earth was of great enough force to save any extraterrestrials. This prevented a planet-hopping Christ, but it strained credulity. In addition, plurality of worlds became a central doctrine for at least two 19th-century religions: the Mormons and Seventh-day Adventists. Yet another religion, the Swedenborgians, had held it as one of their beliefs since the mid-18th century.

Although the idea of other worlds was also incorporated into many secular philosophies, Christianity holds the distinction of being the philosophical worldview that most influenced the plurality-of-worlds doctrine in the 19th century, at least in the Western world. Despite all the discussion, the long-sought resolution between Christianity and pluralism was elusive. Efforts to achieve that resolution would continue in the 20th century, but in a far less dominant role, as science and religion went their separate ways.

1.3 SCIENTIFIC FOUNDATIONS

Two fundamental developments shook science just as the Whewell debate reached its height: in 1859 Charles Darwin (prodded by A. R. Wallace's independent discovery) published his theory of the origin of species and evolution by natural selection, and in the early 1860s the new technique of spectroscopy was applied to astronomy. Though these and less sweeping developments in other fields did not cause an immediate and radical change in the character of the plurality-of-worlds debate, they did signal the beginning of a long-term change that would bring the subject of other worlds increasingly under the purview of science. During this period, Harvard Professor Karl Guthke wrote in *The Last Frontier,* science invested the idea of the plurality of worlds with "a previously unimagined credibility that allows it to penetrate deep into the consciousness of the age."

Much of the progress in the 20th-century debate may be seen as beginning with evolution and astrophysics. Natural selection not only provided the basis for a discussion of the evolution of life under differing conditions beyond the Earth, it also gave impetus to the idea of the physical evolution of the universe. And by examining the chemical fingerprints present in the analysis of starlight, spectroscopy produced a tool to study the nature of the planets and stars in ever-increasing detail. This, in turn, provided for the first time a means for determining the possibility of life on the basis of physical conditions and for demonstrating the idea of cosmic evolution.

Evolution and spectroscopy represented, respectively, breakthroughs in theory and technique. Of the two, spectroscopy would have the more immediate and profound effect on the fate of the biological universe. Though the arguments of analogy and uniformity of nature had long given credence to the belief that the building blocks for matter and life were alike throughout the universe, now for the first time this great truth could be observationally proved. Many of the spectroscopic pioneers themselves did not fail to see the connection of their subject to life in the universe, including Sir William Huggins. Huggins and his colleague William Allen Miller wrote in 1864 that their discoveries provided "an experimental basis on which a conclusion, hitherto but a pure speculation, may rest – viz. that at least the brighter stars are, like our sun, upholding and energising centres of systems of worlds adapted to be the abode of living beings." And Huggins's early attempt to probe planetary atmospheres spectroscopically was the first step toward yet another research program that would be increasingly central to the extraterrestrial life debate in the future. Other pioneers in the new science made similar connections, and their early work and that of their successors put the astrophysical component of the extraterrestrial life debate on a firm footing.

The Darwinian theory of evolution had a more gradual effect, but one eventually no less significant. Its earliest effect was in the support it lent to the idea of the physical evolution of the universe. This evolutionary universe – along with the application of the first results of spectroscopy – is already apparent in one of the most prominent treatises on the plurality of worlds, Richard A. Proctor's *Other Worlds than Ours* (1870). Proctor (Fig. 1.3) conceived his book as a way of bringing the latest astronomical discoveries to the public; it may also be seen as reflecting a trend toward a more scientific approach for the debate. Influenced by teleology, Proctor had originally hoped to show how all the planets were inhabited, but he "found the ground crumbling under my feet. The new evidence . . . was found to oppose fatally . . . the theory I had hoped to establish." This willingness to suspend belief due to the weight of physical evidence was the absolute prerequisite for a scientific approach to the question. But it was also joined by a nascent evolutionary view: while Jupiter, Saturn and the outer planets were probably not now

Fig. 1.3. British astronomer R. A. Proctor.

inhabited, Proctor argued, in the course of their evolution they would one day become as habitable as Venus, Earth, and Mars now were. In *Our Place Among Infinities* and *Science Byways,* both published in 1875, the evolutionary view, in which all planets would attain life in due time, assumed a central role. This evolutionary universe still left Proctor with "millions of millions of suns which people space," of which "millions have orbs circling round them which are at this present time the abode of living creatures." In Proctor the role of teleology was replaced by the eons of time required for evolution.

In France the astronomer Camille Flammarion (Fig. 1.4) assumed Proctor's role of professing to take a scientific approach to the problem of other worlds. It is clear that his *La pluralité des mondes habites* (1862) did not achieve this goal entirely, any more than did Proctor's work. Analogy and plenitude, for example, were mixed with data from the new astronomy in Flammarion's arguments. And if anything, in subsequent publications Flammarion's pluralism became even more radical, taking on the quality of philosophy. Yet, "for all his tendency to enthusiasm and verbosity, Flammarion adheres consistently to this principle of experimental, inductive science," Guthke found. Moreover, at least by the 1872 edition, Flammarion had been

Fig. 1.4. French astronomer Camille Flammarion, at age 20 in 1862, the year of publication of his *Pluralité des mondes*. Flammarion and Proctor (Fig. 1.3) were crucial in turning the debate toward a more modern approach. Flammarion photo courtesy Observatoire de Juvisy. Société Astronomique de France.

deeply influenced by Darwin. Life began by spontaneous generation, evolved via natural selection by adaptation to its environment, and was ruled by survival of the fittest. In this scheme of cosmic evolution, anthropocentrism was banished; the Earth was not unique, and humans were in no sense the highest form of life. Though Flammarion was led to almost religious heights by his pluralism, it was in no sense a Christian view, and was in any case a conclusion reached after the fact rather than an assumption on which his pluralism was based. *La pluralité* reached 33 editions by 1880 and was reprinted until 1921. Through this and numerous other pluralist writings, both fiction and nonfiction, Flammarion would exercise considerable influence on the 20th-century debate, not only through his 19th-century works but more directly through personal enthusiasm until his death in 1925.

By the end of the 19th century, it was the scientific approach of Proctor and Flammarion, stripped of the latter's radical pluralism (not to say evangelical enthusiasm), that ruled the day. The writings of the Irish astronomer Royal Robert S. Ball, the Irish science popularizer J. E. Gore, the American astronomer Simon Newcomb, and many others show much in common with Flammarion and Proctor on the subject. In them Laplace's nebular hypothesis – which not only provided the cosmogonical framework for our own solar system but also pointed to an abundance of planetary systems – became an integral part of the evolutionary approach to nature, with Gore even estimating the probability of planetary systems, a task that would prove a favorite pastime for the 20th century. Further developments, including the application of the kinetic theory of gases to planetary atmospheres, not only added further scientific content to the debate, but also illuminated a later stage of planetary evolution, whereby (based on their masses) it was shown that the Moon could have no atmosphere and Mars no aqueous atmosphere.

The evolutionary view could, however, be carried too far. In the United States, Percival Lowell concluded that because Mars was an older and more highly evolved planet, so were its superhuman inhabitants. The dangers of the scientific approach when applied to such extremely difficult problems as the observation of planetary surface features, the search for planetary systems, and the origin and evolution of life constitute one of the leitmotifs of our study.

A final point needs forcefully to be made. Whatever the scientific merits of the extraterrestrial life debate, the emotional issue of human status is inextricably linked to all discussions of inhabited worlds. Pluralism and anthropocentrism had long been locked in a deadly battle that had not been completely decided by the dawn of the 20th century. Committed anthropocentrists were likely to be the staunchest foes of pluralism, no matter what the evidence, and pluralists – whether they liked it or not – contributed significantly to the demise of anthropocentrism. For all the appeal to scientific argument, the continuing battle between anthropocentrism and other worlds pervades 20th-century discussions of extraterrestrial life and carries the Darwinian debate on the status of humanity into the universe at large. Nowhere is this more evident than in the work of Alfred Russel Wallace (1823–1913), cofounder with Darwin of the theory of evolution by natural selection. His influential work *Man's Place in the Universe: A Study of the Results of Scientific Research in Relation to the Unity or Plurality of Worlds* (1903) incorporated many of the biological problems that would be elaborated in ever more subtle form throughout the century. Although Wallace's book in some ways marks a signal advance in the debate about other worlds, its failure is marked by the dominance of the anthropocentric worldview over all other arguments. Convinced of the nearly central position of the Sun in the universe (Fig. 1.5), Wallace first sought – and found – the significance of this fact in the uniqueness of life, and then adduced arguments in favor of the view that life was

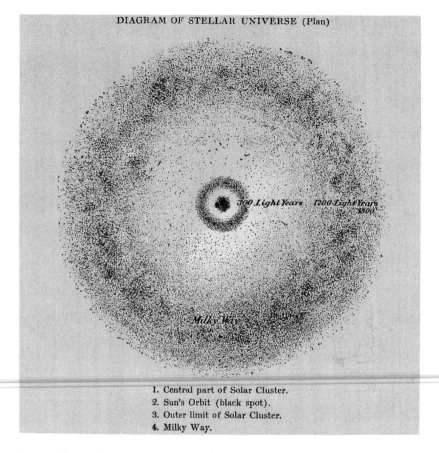

DIAGRAM OF STELLAR UNIVERSE (Plan)

1. Central part of Solar Cluster.
2. Sun's Orbit (black spot).
3. Outer limit of Solar Cluster.
4. Milky Way.

Fig. 1.5. The anthropocentric image of the universe according to A. R. Wallace (1903), showing the Milky Way stellar system 3600 light years in diameter with the Sun (a tiny black dot almost invisible here) near the center.

found beyond the Earth neither in our solar system nor in others. Fifty years after Whewell's treatise, Wallace confidently concluded that "Our position in the material universe is special and probably unique, and . . . it is such as to lend support to the view, held by many great thinkers and writers today, that the supreme end and purpose of this vast universe was the production and development of the living soul in the perishable body of man." Although professing a scientific approach, Wallace's book serves as a lesson on the limits of science when worldviews dominate empirical evidence. It is a lesson the 20th century should take to heart.

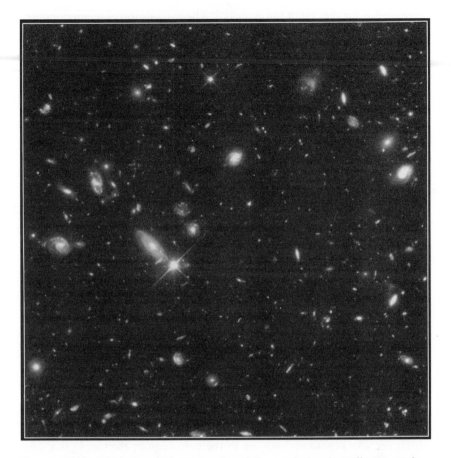

Fig. 1.6. This "Hubble Deep Field" image forcefully reveals that the Milky Way Galaxy is only one of billions of galaxies in the universe. The Hubble Space Telescope image, taken over 10 days in December 1995, is the deepest-ever view of the universe and covers an area of the sky only 1/30th the diameter of the full Moon. It is only the most recent in a series of studies in the 20th century that have shown the noncentrality of our galaxy.

During the 20th century, the tug of war between anthropocentrism and other worlds was profoundly affected by radical changes in the astronomical worldview. While it was still possible as the century began for scientists to argue for an anthropocentric universe based on the Earth's privileged physical position in the cosmos, by 1930 advances in astronomy had destroyed this argument. The resultant worldview – an expanding universe of enormous dimensions in which the solar system was at the periphery of one galaxy among millions (Fig. 1.6) – tipped the scales strongly toward the presumption of other worlds for the rest of the century. "The assumption of mediocrity" became an underlying current of thought favoring other inhabited worlds, superseding the assumption of uniqueness that had opposed it. The hopes for anthropocentrism at the beginning of the century, and its rapid demise thereafter, constitute one of the profound shifts in 20th-century thought. In this sense, the proponents of extraterrestrial life therefore champion not only a scientific theory, but an entire philosophy.

Transcending all problems of observation and methodology is the historical fact that from the time of the ancient Greeks to the 19th century (and even by the time of Kant a century before), the physical world had been transformed into the biological universe, one of the great revolutions of Western thought. By the middle of the 18th century, cosmological worldviews had brought the idea of life beyond the Earth into the mainstream of European consciousness, where it was for a time reconciled with Christianity because its vision of universal life was compatible with an omnipotent God. A century later it had been accepted, reconciled, or rejected by numerous authors, who explored more fully its compatibility not with God, but with Christian dogma. And at the threshold of the 20th century the theory of an evolutionary universe, the development of astrophysics, and refined observational techniques combined to lay the scientific foundations on which the 20th-century extraterrestrial life debate would rest.

By the beginning of the 20th century, the biological universe had achieved the status of a worldview, a "biophysical cosmology" that asserted the importance of both the physical and biological components of the universe. Like all cosmologies, it made a claim about the large-scale nature of the universe: that life is not only a possible implication, but a basic property, of the universe. Like all cosmologies, it redefined our place in the universe. And most important, like other cosmologies, in the 20th century the biophysical cosmology became increasingly testable – even if it still embodied philosophical assumptions along with scientific theory and observation. The remainder of this volume documents the attempts of our century to undertake the exceedingly difficult task of testing the biophysical cosmology and coming to grips with the biological universe.

2

LIFE IN THE SOLAR SYSTEM

Not everybody can see these delicate features at first sight, even when pointed out to them; and to perceive their more minute details takes a trained as well as an acute eye, observing under the best conditions. . . . These are the Martian canals.

Percival Lowell (1906)

The hypothesis of plant life . . . appears still the most satisfactory explanation of the various shades of dark markings and their complex seasonal and secular changes.

Gerard P. Kuiper (1955)

It is impossible to prove that any of the reactions detected by the Viking instruments were not biological in origin. It is equally impossible to prove from any result of the Viking experiments that the rocks seen at the landing sites are not living organisms that happen to look like rocks. Once one abandons Occam's razor the field is open to every fantasy. Centuries of human experience warn us, however, that such an approach is not the way to discover the truth.

Norman H. Horowitz (1977)

Although there are alternative explanations for each of these phenomena taken individually, when they are considered collectively, particularly in view of their spatial association, we conclude that they are evidence for primitive life on early Mars.

David S. McKay et al. (1996)

The Moon having long been considered dead by all but a few eccentrics, it is not surprising that the beginning of the 20th century found the search for life focused on the planets, especially the Earth's nearest neighbors. The search in its earliest form in our century is thus intimately connected with the study of the physical characteristics of the planets, a field that received its theoretical underpinnings from the Copernican theory (which conferred a similar status on Earth and the other planets) and was observationally launched with the invention of the telescope. The importance of the question of life, especially in the case of the enigmatic planet Mars, inspired scientists from Lowell to the Space Age to undertake research on the physical conditions of the planets. The rigid screening of planets for vital conditions narrowed the scope of discussion largely to the planet Mars, though Venus made sporadic appearances on stage early in the century and the intriguing Jovian and Saturnian satellites Europa and Titan at century's end.

The key to progress on the question of life on other planets was therefore the refinement of observational techniques that reveal planetary conditions relevant to life, including temperature, water, and atmosphere. Over the course of the century the increasing power of telescopes and their detectors,

the physical methods of photography, photometry and spectroscopy, and the
ability to send spacecraft to Mars all were brought to bear on the problem of
life in the solar system. And when meteorites of Martian origin were identified on the surface of the Earth at the end of the century, only sophisticated
analytical laboratory techniques allowed the claims of Martian microfossils
to be credible. Technique and scientific conclusions are therefore closely in-
terwoven in the debate over extraterrestrial life.

Despite all the sophisticated techniques, the extremely contentious arguments hinging on planetary observations reveal the problematic nature of observation in science, whether undertaken from the surface of the Earth or the
surface of Mars. Yet – and this is a crucial point – despite the many problems of observation, the debate progressed throughout the century from the
question of intelligence in the solar system, to vegetation, and finally to microorganisms, organic molecules, and fossil life. False leads, emotional debate, and wishful thinking notwithstanding, the historical record shows that
20th-century science did narrow the question of life in the solar system to
fossil or microbial life on Mars or a few satellites of the giant gas planets.
Precisely how science reached this momentous conclusion, through a thicket
of subtle problems, is the subject of the present chapter.

2.1 LOWELL AND MARS: THE SEARCH FOR INTELLIGENCE, 1894–1924

The central character in the search for intelligence on Mars, surely one of
the strangest and most contentious episodes in the history of science, was
the unlikely figure of Percival Lowell (Fig. 2.1). A wealthy Bostonian businessman with a science and mathematics background from Harvard, Lowell
neither initiated nor concluded the debate about "canals" on which his theory
of Martian intelligence depended. But for 23 years, from 1894 to his death
in 1916, he was at once its leader and most eloquent spokesman, and thus
the eye of the storm that raged around the question of intelligence on Mars.
At the height of the controversy around 1910, his name was known to scientists and laymen alike, in America and throughout Europe, where his writings
were avidly read and his lectures invariably popular. Two primary questions
cry out for an answer in connection with the controversy: How did Lowell
arrive at his startling theory that rectilinear features of Mars were proof that
the planet was inhabited? And how was it that science was unable to put this
theory to rest for almost two decades?

The outlines of the story, its key participants, and several crucial characteristics of the debate, are evident in Table 2.1. We see here that the canal
controversy began with the Italian astronomer Giovanni Schiaparelli in 1877,
was given impetus by the confirmation of Henri Perrotin and Louis Thollon

Fig. 2.1. Lowell at the 24-inch refracting telescope purchased from Alvan Clark in 1896, with which he made many observations of Mars, often with decreased aperture. In this view, Lowell is observing Venus by daylight. Lowell Observatory photograph.

Table 2.1. *Observational highlights of Martian canals, 1877–1924*

Observer	Date	Observatory	Instrument	Method	Result/remarks
Schiaparelli	1877	Brera (Milan)	8 in.	Visual	First observation of system of *canali*
Green	1877	Madeira	13-in. reflector	Visual	Shaded area boundaries; no canals
Hall/ Harkness	1877	U.S. Naval	26 in.	Visual	Moons of Mars found but no canals
Schiaparelli	1879/ 1880	Brera (Italy)	8 in.	Visual	First report of double canals
Maunder	1882	Greenwich	28 in.	Visual	Some canals
Perrotin/ Thollon	1886 1888	Nice (France)	15 in. 30 in.	Visual	Many canals; first confirmation of double canals
Holden/ Keeler et al.	1888	Lick	36 in.	Visual	Some canals but no doubles
Pickering/ Douglass	1892	Harvard Arequippa (Peru)	13 in.	Visual	Canals and "lakes" at canal junctions
Barnard	1892	Lick	36 in.	Visual	Some canals but not fine lines
Lowell/ Pickering/ Douglass	1894 1895	Lowell	12 in. 18 in.	Visual	Canals artificial
Antoniadi	1894/ 1896	Juvisy (France)	9.6 in.	Visual	42 canals; 1 double doubling is illusion
Cerulli	1896	Teramo (Italy)	15.5 in.	Visual	Illusion – optical origin
Lampland	1905	Lowell	24 in.	Photographic	Canals photographed
Todd/ E. Slipher	1907	Chile	18 in.	Photographic	Canals photographed
Antoniadi	1909	Meudon	33 in.	Visual	Canals resolved
Hale	1909	Mount Wilson	60-in. reflector	Visual	Much detail, no canals
Trumpler	1924	Lick	36 in.	Visual/ photographic	Strips of vegetation

Sources of data: M. J. Crowe, *The Extraterrestrial Life Debate* (Cambridge, 1986); W. G. Hoyt, *Lowell and Mars* (Tucson, Ariz., 1976), and primary sources. Instruments are refractors unless specified otherwise.

in France in 1886 and 1888, took on its sensational aspect with Lowell's theory in the United States in 1894, and saw the beginnings of final resolution with Eugene M. Antoniadi's observations in France in 1909. We note too that the initial observations in 1877 were made with a very modest telescope, that not everyone saw canals even with larger telescopes, and that those who did see them did not always see them in exactly the same way. Moreover, Table 2.1 reveals that most of the controversy centered on visual observations made by astronomers at the eyepiece of the telescope, though by 1905 photographic confirmation was believed to be at hand. Finally, the table illustrates that the debate was carried out not by a scientific subculture, but by leading astronomers representing reputable institutions around the world. What Table 2.1 does not show are the hundreds of lesser participants in the debate and the press hype that made Mars a public and scientific cause célèbre. Nor does it show the ancillary arguments over planetary atmospheres and water that were central to that debate.

The approximately 15-year intervals between Schiaparelli's detection of a Martian canal network and the elaboration of Lowell's theory, and again between Lowell's theory and the resolution of the canals question, are no coincidence. Indeed, these intervals reflect a basic astronomical fact that is also the cause for the ebb and flow in Martian research graphically apparent in Figure 2.2: approximately every 2 years the distance between Earth and Mars is minimized (this is termed an "opposition" because Mars is opposite the Sun in the sky), and approximately every 15 years they are even closer. As a result, the apparent size of the disk of Mars viewed from Earth varies enormously; when we consider that even during oppositions on average Mars is only 1/100th the diameter of the full Moon, we begin to see the difficulties that Martian observations faced from the beginning. The year 1877 marked one of the 15-year close approaches of Mars, and that year saw not only Schiaparelli's first observation of an extensive system of canals, but also Asaph Hall's discovery at the U.S. Naval Observatory of the two moons of Mars. In addition, the altitude of the planet was a factor in observing it; the elaborate dance between altitude and distance, graphically evident in Figure 2.3, determined the best observing conditions. Even under the best conditions, however, one still had to contend with viewing through the Earth's turbulent atmosphere, as with all astronomical observations.

Genesis of the Theory

How then did Lowell arrive at his theory? It is clear that he was neither the first to suggest intelligence on Mars nor the first to observe the canal phenomenon upon which he based his theory. Schiaparelli himself provided ample stimulation. In an important lengthy article on the eve of the 1894 opposition,

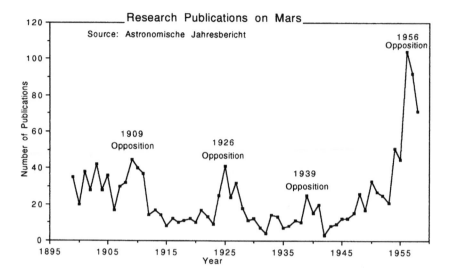

Fig. 2.2. Research publications on Mars, 1900–1957. Publications increased with favorable approaches of Mars and increased in general as the Space Age approached. Based on data from the annual volumes of *Astronomische Jahresbericht* (1900–1957).

Schiaparelli described Mars as a planet with two polar caps composed of snow and ice, seas and continents arranged very differently from those on Earth, and an atmosphere rich in water vapor. It was also a planet of change, for the melting polar caps seemed to produce a temporary sea around the northern cap, and Schiaparelli believed this water was distributed over great distances by "a network of canals, perhaps constituting the principal mechanism (if not the only one) by which water (and with it organic life) may be diffused over the arid surface of the planet." From their similar appearance to seas, and from the phenomenon of the melting snows, Schiaparelli concluded that the canals constituted "a true hydrographic system," but "notwithstanding the almost geometrical appearance of all of their system, we are now inclined to believe them to be produced by the evolution of the planet, just as on the Earth we have the English Channel and the Channel of Mozambique." Still, he left the door open to the artificial hypothesis: "I am very careful not to combat this supposition, which includes nothing impossible." Though in the end he concluded that nature itself can produce geometric regularity such as crystalline forms, his essay was very suggestive.

The first piece of the puzzle in understanding the genesis of Lowell's theory, then, is that such speculations were in the air, and particularly in Schiaparelli's essay, which appeared in Italian in February 1893 and was translated soon

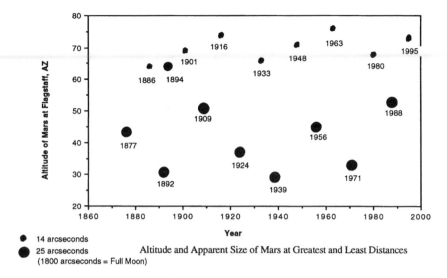

14 arcseconds

25 arcseconds   Altitude and Apparent Size of Mars at Greatest and Least Distances

(1800 arcseconds = Full Moon)

Fig. 2.3. As Mars approaches and recedes, its apparent size changes as viewed from Earth, from about 25 arcseconds to 14 arcseconds during oppositions and as small as 4 arcseconds at its most distant. The full Moon is 1800 arcseconds by comparison. With the exception of 1894, only maximum and minimum opposition data are graphed here. The relatively large disks in 1877, 1892, and 1909 made those important years for observing Mars. The low altitude of Mars in 1892 spurred the Harvard Southern Hemisphere observations of W. H. Pickering in Peru. The opposition following the closest approach finds the planet much higher in the sky but the size of the disk already reduced by 20 percent. Lowell began his observations at Flagstaff in 1894 at such a time. Altitudes are plotted here for Flagstaff; Schiaparelli's observatory in Milan was 10 degrees farther north, so all the altitudes shown here would have been lower by 10 degrees in the sky. Antoniadi would have viewed Mars almost 14 degrees lower than observers at Flagstaff.

thereafter for Lowell. As Lowell entered the debate, the general opinion was that the canals were cracks in the Martian crust made during the planet's solidification, but some argued for their artificiality.

But what precipitated Lowell's entry into the debate? Was he responding directly to Schiaparelli's challenge or to some more deeply rooted interest? Why did Lowell, a businessman and traveler with a penchant for the exotic, enter astronomy at all? Born into a wealthy family in 1855, Lowell was attracted to astronomy from boyhood, when he scanned the heavens with a 2-inch telescope. He attended Harvard from 1872 to 1876 and studied with the famous mathematician Benjamin Peirce, who described him as one of his most brilliant students. Another Harvard influence may have been John Fiske

(1842–1901), who presented his evolutionary cosmic philosophy in *Outlines of a Cosmic Philosophy Based on the Doctrine of Evolution* in 1874, 2 years before Lowell graduated. That Lowell embraced such a philosophy is evident from his commencement address on "The Nebular Hypothesis." But until 1893 he spent his time managing family business affairs and traveling, the latter in which he achieved fame as an author of books on the Far East.

The year 1893, and the appearance of Schiaparelli's provocative article, found Lowell still in the Far East. In the autumn of that year he returned to Boston; within a few months, the *Boston Herald* reported that he was financing an expedition to Arizona in conjunction with Harvard University, to be headed by Harvard astronomer William H. Pickering. The Harvard connection was soon severed, but Pickering, who had observed lakes and clouds on Mars, remained a real influence. A Christmas gift to Lowell in that year of Flammarion's *La planète Mars* clearly showed Schiaparelli's canals. The picture that emerges from the scattered evidence is that Lowell certainly knew of Schiaparelli's work and perhaps his failing eyesight, that in a more immediate sense Pickering played a crucial role in bringing Mars and its problems to the attention of Lowell, and that it was Pickering who first suggested the Arizona site, a long-cherished idea of his. It was in mid-January 1894 that Lowell made his offer to finance the expedition; Pickering immediately began to assemble the necessary equipment with the help of the telescope maker Alvan Clark, and Pickering's assistant A. E. Douglass was assigned the task of site testing.

What began as a Harvard expedition financed by Lowell ended as a new observatory where the search for life on Mars was clearly the goal. In a paper read before the Boston Scientific Society in May 1894, Lowell noted that the observatory's work was to be "an investigation into the condition of life in other worlds, including last, but not least, their habitability by beings like (or) unlike man. This is not the chimerical search some may suppose. On the contrary, there is strong reason to believe that we are on the eve of [a] pretty definite discovery in the matter." Lowell went on to make it clear that the source of this belief was a nonanthropocentric worldview and the nebular hypothesis; given these, "then to develop life more or less distantly resembling our own must be the destiny of every member of the solar family which is not prevented by purely physical considerations, size and so forth, from doing so." With an early interest in astronomy, a taste for the exotic, a knowledge of the work of Schiaparelli on Mars, intrigued by the notion of that planet as part of an evolutionary scheme that might include life, and with a direct push from W. H. Pickering, Lowell's subsequent actions begin to make more sense.

To a remarkable extent, Lowell's works are different from other treatises on the plurality of worlds. No Proctor or young Flammarion here, enthusiastically assessing, mostly secondhand, the chances of life in the universe. Not even a more sober and scholarly Flammarion, concentrating on Mars to compile observations and assess "its conditions of habitability." Rather, Lowell's books and articles are reports of observations that he himself largely made or sponsored, culminating in his first work with the claim that Mars is not only habitable, but actually inhabited. Not that Lowell should be accused of being provincially Martian; difficult as the observations were, Mars was simply the closest planet that opened the door to more general claims. While others sought a general treatment heavily laced with philosophical arguments, Lowell's three major works – *Mars* (1895), *Mars and its Canals* (1907), and *Mars as the Abode of Life* (1908) – as well as three volumes of annals issued from the Lowell Observatory during his lifetime and a considerable number of its 73 *Bulletins,* concentrated all of his considerable energies on Mars as a case study of the more general belief. In this may be found his unique contribution, his drive, and the source of his reputation as a man of controversy.

Though with his financial means Lowell could easily construct an observatory, it would not seem to be so easy to construct and sustain a theory unless there was a considerable observational basis for doing so. Yet Lowell's first paper on Mars appeared in August 1894, and already there his theory was well advanced. After describing his observations of canals but not clouds Lowell wrote, "Here we have a *raison d'etre* for the canals. In the absence of spring rains a system of irrigation seems an absolute necessity for Mars if the planet is to support any life upon its great continental areas." By the time of his first book on the subject the following year, that theory was set forth in full. Perhaps the speed of the theory's formation is itself an indictment.

Lowell's 1895 volume leaves no doubt as to his "chain of reasoning" from observations to intelligence, at least in retrospect:

> To review, now, the chain of reasoning by which we have been led to regard it probable that upon the surface of Mars we see the effects of local intelligence. We find, in the first place, that the broad physical conditions of the planet are not antagonistic to some form of life; secondly, that there is an apparent dearth of water upon the planet's surface, and therefore, if beings of sufficient intelligence inhabited it, they would have to resort to irrigation to support life; thirdly, that there turns out to be a network of markings covering the disk precisely counterparting what a system of irrigation would look like; and, lastly, that there is a set of spots placed where we should expect to find the lands thus artificially fertilized, and behaving as such constructed oases should. All this, of course, may be a set of coincidences, signifying nothing; but the probability points the other way.

It is clear from this passage that the conditions for life were crucial, for if life was impossible, Lowell's theory of artificial canals was impossible. In this connection Lowell argued that the long-known fact of changes in the polar caps, and Schiaparelli's recent detection of changing tints on the Martian surface, were proof of an atmosphere, for a planet's atmosphere is the agent of change. He further pointed to measurement of variations in the diameter of the planet and even inferred from the kinetic theory of gases that its molecules must be similar to those of the Earth. As to surface water, Lowell argued that the polar cap was composed of aqueous snow, as opposed to frozen carbon dioxide, and that a polar sea was produced after the cap melted in the summer. In his discussion of "areography" Lowell argued that the blue-green areas, considered to be seas by some, were actually areas of vegetation that come to life with the melting of the polar caps, so that, far from open seas in abundance on the planet, water was actually very scarce, just as one would expect for a planet more evolved than Earth. Any inhabitants on the planet would therefore need a system of irrigation – and this is just what is seen. Not only did Lowell find Martian conditions conducive to life, he also found in those conditions an explanation for the canals.

In the end Lowell and his staff at Flagstaff catalogued 183 canals, four times the number Schiaparelli had seen (Fig. 2.4). The theory of their artificiality Lowell built on their "supernaturally regular appearance," based on their straightness, uniform width, and systematic radiation from special points. He had no illusions that he was seeing the canals themselves; instead, he believed he was seeing the strips of fertilized land bordering them, an explanation he credited to William Pickering. And therein lay for him further proof of their nature. For Lowell noted that the canals change appearance, growing progressively from the pole toward the equator as the polar caps melt. This does not happen suddenly, but with a delay that might be seen if vegetation needed time to renew and make itself visible. And the oases were seen as further evidence that the canals are constructed so as to fertilize them.

By 1895, then, Lowell's theory of artificial canals on Mars was fully set forth, not in the cautious and brief manner of Schiaparelli 2 years earlier, but in an ebullient and highly readable book that threw caution to the wind. Thrust into the open, an issue full of public interest, a golden opportunity now presented itself for science either to confirm an earthshaking theory or to crush it quickly under the glare of objective argument. To the dismay of the public and the embarrassment of astronomers, science was unable to do either for almost two (some would say seven) decades.

Resolution of the Conflict

Why, despite their best attempts, did astronomers take so long to resolve the issue of the canals of Mars? The answer lies not only in the disputed appearance of the canals, but also in disputed ideas of the proper route from observation to theory and in disputed observations of planetary conditions that, had they been more definitive, might have settled the controversy themselves by precluding intelligence. Far from embodying the objectivity that the public expected from science, over the next two decades and more the opinions of scientists on the Martian canals would cover the entire spectrum of possibility, from those who denied even the reality of the canals, on the one hand, to those who accepted Lowell's theory of artificial Martian constructions, on the other.

Among the most illuminating immediate reactions to Lowell's theory was Lick astronomer W. W. Campbell's review of *Mars*, which he penned for the respected journal *Science*. Although Campbell accepted the wave of darkening and the organic origin of these dark areas, on Lowell's more breathtaking journey from observation to theory he was insistent: "Mr. Lowell went direct from the lecture-hall to his observatory in Arizona; and how well his observations established his preobservational views is told in his book." While Campbell accepts the canals as real surface features, he ridicules the artificial canal theory as impossible from a hydraulic point of view, and no more of intelligent origin than the long, straight markings on the Moon that radiate from the crater Tycho. Campbell did not argue that conditions on Mars were unsuitable for intelligent life; rather, he stated that Lowell's view of Mars's atmosphere was in accord with his spectroscopic results of 1894, failing to detect water vapor at a certain level. In short, Campbell objected to some of Lowell's observations and certainly to Lowell's pathway from observation to the theory of artificiality, but he was unable to undermine completely Lowell's position by appeal to impossible conditions on the planet. For all of these reasons, the case against Lowell was inconclusive.

As the controversy picked up steam, the combatants became so numerous, and their opinions so varied, that it is easy in retrospect to lose sight of the trajectory of the argument. At the conservative end of the spectrum were the illusionist arguments, the champion of which was E. W. Maunder (1851–1928), astronomer at the Royal Greenwich Observatory near London. Already in 1894 Maunder had applied his work on sunspot groups to the canals of Mars issue and concluded that they were "the summation of a complexity of detail far too minute to be ever separately discerned," a position he bolstered over the next decade. But those at the other end of the spectrum, including more than just Lowell, were not impressed. The young Henry Norris Russell, destined to become the dean of American astronomers, wrote of the canals as late as 1901, "Perhaps the best of the existing theories, and certainly the most stimulating to the imagination, is that proposed by Mr. Lowell and his fellow workers at his observatory in Arizona." And although by 1905 Lowell claimed that his staff had actually photographed Martian canals, the fact was that photography had only succeeded in showing a few canals, not

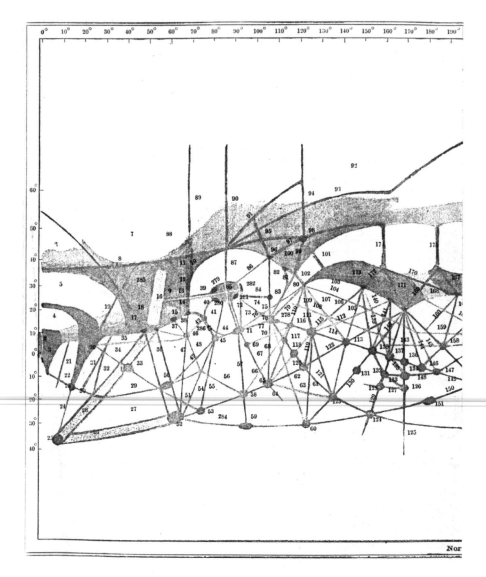

Fig. 2.4. Lowell's first map of Mars (1895).

an entire network, and even if one accepted the few canals photographed, the growing camp of illusionists still questioned the reality of the vast majority. Between Lowell and the illusionists was a group that argued that the canals were real but natural formations composed of cracks in the Martian surface.

36

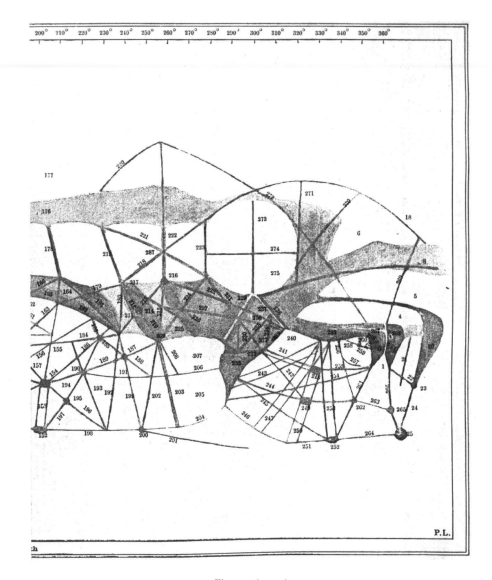

Fig. 2.4 *(cont.)*

In a hundred variations, these arguments were echoed around the scientific world and reechoed and amplified in the popular press.

If the interpretations of canal observations were indecisive, the study of the Martian atmosphere and surface conditions held the promise of excluding

Fig. 2.5. E. M. Antoniadi. Courtesy Royal Astronomical Society; photo by Phébus, Constantinople.

the possibility of life by virtue of the planet's harsh environment. In 1908 V. M. Slipher reported from the Lowell Observatory the observation of water vapor in the spectrum of Mars, igniting that controversy once again. Not to be outdone, during the 1909 opposition, Campbell mounted an expedition to observe the same phenomenon from the top of Mt. Whitney, and concluded that water vapor was very scarce on Mars even compared to the amount above Mt. Whitney.

These disparate conclusions left the conditions for life on Mars still uncertain. And so, 15 years after Lowell had first published his hypothesis, astronomers worldwide prepared once again to observe the surface features of Mars under the most favorable conditions since 1894. Contemporary astronomers and modern historians agree that it was Eugene M. Antoniadi (1870–1944; Fig. 2.5), a Greek-born astronomer known later in life for his studies on ancient Greek and Egyptian astronomy, whose observations contributed most to solving the riddle of the canals of Mars.

It is difficult to escape a number of ironies in Antoniadi's clinching role: that it was in Flammarion's France, hotbed of pluralism, that Lowell's thesis would meet its Waterloo; that Antoniadi himself had been inspired by the great pluralist Flammarion and was actually working at Flammarion's observatory at Juvisy in 1893–1894 as Lowell began his entry into astronomy; and that many of Antoniadi's results would be published in the journal of the French astronomical society that Flammarion founded. Above all, there is the apparently anomalous fact that Antoniadi solved the riddle using a 33-inch refractor – smaller than the 36-inch refractor available at Lick Observatory in 1888. Why, we may well ask, did not the Lick astronomers resolve the canals into their constituent parts in 1888 (or Yerkes astronomers with the 40-inch refractor after 1897) and spare the world the entire Lowell controversy? For that matter, why did the Meudon 33-inch telescope, the largest in Europe, not resolve the canals shortly after its erection in 1891?

A closer look at Antoniadi does not entirely explain the anomaly. Although as late as 1903 he still believed in "the incontestable reality" of many of the canals, when given the use of the 33-inch refractor at Meudon in 1909, he approached these observations with "an open mind." When 36 canals were examined "steadily under good seeing," all corresponded to something "real," such as diffuse streaks or borders of shaded areas, but not "canals." Other so-called canals, seen only in "flashes" lasting about one-third of a second, corresponded to nothing on the surface. But the geometrical canal network, he concluded, was "an optical illusion." By December 1909, Maunder rejoiced that scientists and the public "need not occupy their minds with the idea that there were miraculous engineers at work on Mars, and they might sleep quietly in their beds without fear of invasion by the Martians after the fashion that Mr. H. G. Wells had so vividly described."

Antoniadi's results received quick support from the United States, where the era of reflectors had dawned and the new 60-inch telescope at Mt. Wilson had come into use in 1908. George Elery Hale, founder and Director of the Mt. Wilson Observatory, wrote to Antoniadi in January 1910 to report the accord of Antoniadi's visual drawings with observations of the 60-inch telescope. "I am thus inclined to agree with you in your opinion (which coincides with that of Newcomb) that the so-called 'canals' of Schiaparelli are made up of small irregular dark regions." Citing other observations from large telescopes, Antoniadi concluded that "the frail testimony of small refractors has vanished before the decisive evidence of giant instruments; and the telescopes of Princeton, Lick, Yerkes, Mount Wilson, and Meudon have settled the question for ever."

After 1909 Antoniadi continued his observations and solidified his claim of resolved canals. Not surprisingly, Lowell argued valiantly against Antoniadi right up to his death in 1916, stating that Antoniadi was seeing the blurring of continuous lines. But the handwriting was on the wall for canals, and everywhere astronomers hailed the observations of Antoniadi and Hale. A lifetime of work on the subject was summed up in Antoniadi's magnum opus *La planète Mars* (1930), the capstone to the visual era in the canals-of-Mars

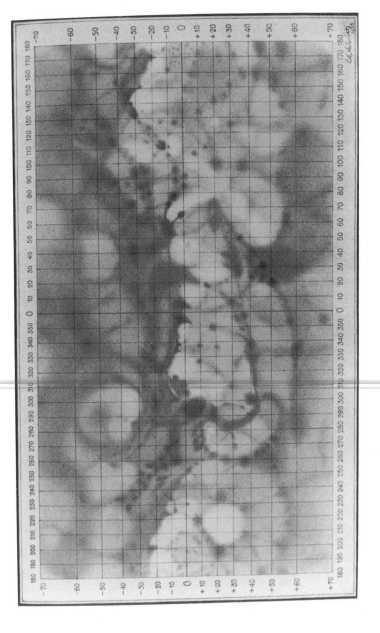

Fig. 2.6. E. M. Antoniadi's map of Mars, showing dark patches but not the fine lines interpreted by Lowell as canals. From Antoniadi, *La planète Mars* (1930).

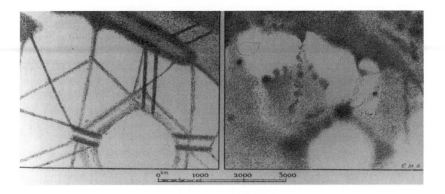

Fig. 2.7. Antoniadi's comparison of his Mars observations with those of Schiaparelli, showing how dark splotches were interpreted by Lowell as fine lines. Schiaparelli's observations were made with telescopes much smaller than the 33-inch one Antoniadi used. From Antoniadi, *La planète Mars* (1930).

controversy, initiated with Schiaparelli's 1878 *Osservazione* and punctuated with Lowell's 1895 *Mars*. Here Antoniadi gave his state-of-the-art version of Martian topography (Fig. 2.6) and repeated his belief that only through large telescopes could we finally understand the illusion of canals, even if they had some basis in streaks and broken spots (Fig. 2.7). Despite this conclusion, Antoniadi believed that Mars possessed an atmosphere because the growing and shrinking polar caps required at least a thin atmospheric blanket. He also realized that Campbell's spectroscopic results indicated a desertlike Mars, a world "in a state of advanced decrepitude." But at the same time, he joined in accepting what was already becoming the new view of Mars – that life is probable on that planet, though in the form of flora, not fauna.

In retrospect we may say that the resolution of the canals-of-Mars controversy was slow in coming for several reasons. The incredible distance of Mars, the Earth's churning atmosphere, the telescope itself, and the bias of the individuals all combined to prolong the debate. After Lowell's death, W. W. Campbell concluded that too many astronomers had tried to solve a problem beyond the limits of the science of the time. With the hindsight of history there is no doubt that the instruments of the time were inadequate to the task before them, and that more than the normal latitude was left for individual error in the form of personal equation. They were right even more than they knew, for history has now provided us with the "truth" about the canals. With few exceptions, we know that Lowell's canals correspond to no actual surface features on Mars (Fig. 2.8). Not only was Lowell wrong, so was Antoniadi in the sense that he believed he had resolved the canals.

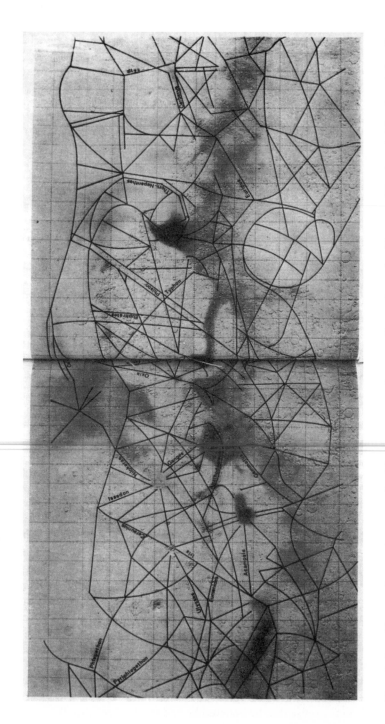

Fig. 2.8. Lowellian canal network compared to Mariner cartography, according to the study of Sagan and Fox (1975). They found virtually no matches between Lowell's canals and real surface features. From *Icarus*, 254 (1975), 610, by permission of Academic Press and the authors.

But the problem of the canals was not entirely one of observational fact; beyond inadequate instruments and personal equation there was the problem of turning this observational fact into evidence for a theory. Moreover, what stands out is that Lowell was a scientist with ample imagination. Nor was he alone in drawing imaginative conclusions in the canals debate or in clinging to those conclusions. In this respect, the canals-of-Mars controversy points toward at least two extreme "cultures" of scientists: one concentrating on gathering the hard facts and unwilling to go beyond them, the other seeing the hard facts as a basis for a larger theory, with all of the imaginative leaps that this implies. These are undoubtedly the extreme ends of an entire spectrum of scientific types. And if this is so, surely Lowell and his allies and opponents were neither the first representatives of the two scientific cultures nor the last.

It is also well to remember, in connection with scientific inference, imagination, and Lowell's reputation, that for a while in the post-Lowellian era the planet Venus was viewed as a possible abode of life, even intelligence. Although it had nothing to do with Lowell (even though he had reported very broad streaky canals on Venus), and although the discussion never reached the fever pitch of the one concerning Mars, the protagonists were prominent scientists. One, the Nobel Prize winner Svante Arrhenius, argued in 1918 for a lush, steamy Venus with low life forms. The other, the distinguished director of the Smithsonian Astrophysical Observatory, Charles Greeley Abbott, argued that cloud-shrouded Venus was the most Earth-like planet and might even harbor intelligent life. Unlike Mars, however, Venus quickly fell victim in the extraterrestrial life debate to harsh conditions. It was considered unlikely by 1930 with the failure to detect oxygen, was held even less likely by 1940 with the detection of abundant carbon dioxide and the realization of a possible greenhouse effect, and was widely seen to be impossible by 1960 with the determination of surface temperatures approaching 480°C. Venus thus provides an interesting "control experiment" of scientific attitudes toward extraterrestrial life when (unlike Mars) evidence accumulated that the physical conditions on the planet were so extreme that life by anyone's definition would be impossible.

Despite the abandonment of Venus as an abode of life, Mars remained a subject of lively debate even after the demise of Lowellian canals. With Schiaparelli dead since 1910, Lowell since 1916, and Flammarion in the last year of his life, science was set for a new generation of Martian investigations during the favorable oppositions of 1924 and 1926. Though Antoniadi's 1930 treatise gave no indication of it, in the search for vegetative life on Mars new astrophysical methods were supplementing the visual topographic tradition, dominant in Martian studies for 50 years, of which Antoniadi represents the apex.

2.2 THE SEARCH FOR MARTIAN VEGETATION, 1924–1957

The extent to which hopes for life on Mars were alive by the late 1920s was well summarized for the public in a symposium published in the *New York Times Magazine* for 1928. Subtitled "Eminent Astronomers Give Their Reasons for Belief That Life Exists on the Great Red Planet," the article spoke of the "traditional theory that Mars is dead" and held that scientists skeptical of Martian life a few years before had now reached some consensus that life was part of the riddle of Mars after all. But the contrast to the Lowellian period was sharp: of the scientists polled, only W. H. Pickering believed it to be "almost certain" that intelligence existed on Mars. C. G. Abbot, director of the Smithsonian Astrophysical Observatory, held the opposite belief: that physical conditions limited Martian life to very low forms of vegetation. Most of the other astronomers, including Henry Norris Russell at Princeton, Harlow Shapley at Harvard, William Coblentz at the National Bureau of Standards, Walter S. Adams at Mt. Wilson, E. B. Frost at Yerkes, and E. C. Slipher at Lowell, affirmed a belief in vegetation and admitted the possibility of low types of animal life.

While the Lowellian theory of a geometric, artificially constructed network was largely rejected by this time, the canals themselves, viewed as natural features, were still very much a part of the new period of Martian studies. The persistent ambiguity about the canals is evident in Russell's viewpoint in the 1928 *New York Times Magazine* article: "At the present time it is generally recognized that there exists an objective basis for the canals in the form of fine detail on the surface of Mars, and it is widely believed that these details have, in a general way, the streaky character of the canals. But the existence of a geometrical network is doubted or denied by a large majority of astronomers. It is therefore necessary to render a verdict not proved with regard to this theory." Even if one accepted the theory of Maunder and Antoniadi that continuous canals were an illusion caused by disconnected patches, the nature of the patches and their apparent rectilinear alignment remained a mystery.

But the remarkable and novel feature of the period beginning with the 1924 opposition is that the story of life on Mars moved beyond the canals to the physics of the planet, from areography to areophysics, as the astronomer Gerard de Vaucouleurs later put it. "Physical methods were first applied on a large scale to the study of Mars during the perihelic oppositions of 1924 and 1926," he wrote in 1954, and in so doing he perceptively delineated a new era in Martian studies. To be sure, the pioneering methods of astrophysics had been applied earlier to Mars; we need only recall Campbell's work on Martian water vapor in 1894 and 1909, among others. But by the early 1920s the field was maturing, instrumentation was being refined, and promising new techniques were being brought to bear on a wide variety of questions related to the physical conditions on Mars: its temperature, its atmospheric pressure and composition, and its changing surface features. Those features included possibly aqueous polar caps, tantalizing indications of vegetation in the form of seasonal changes, and the "wave of darkening" as the polar caps melted, as well as continued interest in the nature of the infamous canals. The move from visual observations to astrophysical techniques also ushered in a new era in the study of Martian biology, an era when the search for intelligence gave way to the search for vegetation, bringing new possibilities for resolution of an age-old question but – alas – also a whole new set of observational problems.

As the favorable oppositions of 1924 and 1926 approached, Mars was believed to have an average temperature well below the freezing point of water, a thin atmosphere of unknown composition, and an enigmatic surface subject to change. As in the past, those enigmatic surface features were scrutinized again. Especially notable were the conclusions of Robert Trumpler's detailed study of Mars undertaken at Lick Observatory, which indicated to him that, while the canals are the result of natural topography, vegetation caused the dark areas and made the canals visible. The observations of E. C. Slipher at Lowell and W. H. Wright at Lick also strengthened the hypothesis of Martian vegetation. At least a part of Lowell's legacy was thus still alive with the close approach of Mars in 1924. Summarizing the visual and photographic evidence Slipher argued that the shifting dark markings of the planet "all obey the law of change that we should expect of vegetation."

But what was new with the 1924 opposition of Mars was the possibility, made real by technical advances in physical methods, of more accurately determining Martian conditions. The new experiments, the highlights of which are given in Table 2.2, now sought to confirm, quantify, or extend the visual and photographic results. Central to this new era was the work of two independent teams attempting to measure the temperature of Mars. Both William W. Coblentz and C. O. Lampland at Lowell Observatory, and Edison Petit and Seth Nicholson at Mt. Wilson, used the newly invented vacuum thermocouple for their experimental determinations. Both teams came to similar conclusions about the temperatures on Mars, namely, that they were cold but sometimes rose above freezing. But only one of the four scientists extensively discussed the implications of his results for the question of life on Mars. In the popular article "Climatic Conditions on Mars," published in 1925, Coblentz speculated that "the observed high surface temperatures on the dark areas of Mars may be explained on the basis of the presence of living vegetation superposed upon a dry vegetable mold which is a non-conductor of heat." With practically no water on Mars, Coblentz reasoned, "the decay and disintegration of vegetable matter would be slow and there would be a

Table 2.2. *Milestones in Martian observations related to life, 1924–1957*

Observer	Date	Location	Instru- ment	Method	Result
Temperature					
Coblentz/ Lampland	1924	Lowell	40-in. reflector	Thermocouple	−28°C for whole disk −10°C to 5°C for bright 10°C to 20°C for dark
Pettit/ Nicholson	1924	Mount Wilson	100-in. reflector	Thermocouple	−13°C disk by first method −33°C disk by second method
Atmospheric pressure and composition (Earth = 1000 mb)					
Adams and St. John	1925	Mount Wilson	60-in. reflector	Prism spectrograph	H_2O 6% of Mount Wilson O_2 16% of Mount Wilson
Menzel	1926			Visual and photographic albedo	Pressure < 5 cm Hg
Lyot	1929	Meudon	33 in.	Visual polarimetry	Pressure < 1.8 cm Hg
Adams and Dunham	1933	Mount Wilson	100-in. reflector	Grating spectrograph	No oxygen No water vapor
Barabashoff	1934	Kharkov	8 in.	Photographic photometry	50 mb
DeVaucoleurs	1945	Perdier		Visual photometry	93 mb
Kuiper	1947	McDonald	82-in. reflector	Infrared (IR) spectroscopy	CO_2
Hess	1948	Theoretical		Meteorology	80 mb
Dollfus	1948– 1951	Pic du Midi	24 in.	Visual polarimetry	83 mb
Surface features/direct indications of life					
Slipher	1924	Lowell	24 in.	Spectroscopy	No chlorophyll
Trumpler	1924	Lick	36 in.	Visual/ photographic	Canals, but not artificial
Pettit	1939			Visual	Canals
E. Slipher	1927– 1962		24 in.	Photographic	Canals, vegetation
Dollfus	1946– 1948		24 in.	Polarimetry	Clouds
Tikhov	1947– 1949			Laboratory studies of reflection spectra	Vegetation
Sinton	1957	Harvard	61-in. reflector	IR spectroscopy	Organic molecules vegetation

Instruments are refractors unless otherwise specified.

slow accumulation of the dry matter of the preceding season which would protect the living plants from the extreme cold of winter."

With conditions similar to north temperate and frigid zones of the Earth, except for atmosphere, Coblentz compared Mars with Siberian conditions, where moss and lichen tundra existed, as well as color changes with season. In addition to temperature, Coblentz noted, visual observations of coloring and darkening in the southern hemisphere of Mars by Lowell, Pickering, Slipher, and others were "strikingly similar" to the observations of areas like Siberia near the terrestrial polar regions. Because certain types of moss have the capacity to raise the temperature above a barren surface due to their high absorption of solar radiation, the presence of lichen and moss would explain the high temperatures. But, Coblentz cautioned, entirely different types of plants might have evolved on Mars due to environmental conditions. While vegetable and perhaps animal life is possible on Mars, any animal life that cannot migrate "must be trogdolytic, able to burrow deep and hibernate, or able to withstand the long sieges of intense cold in a benumbed state, as do, for example, the torpid insects which one finds on warm days in winter."

The combination of physical radiometric measurements with visual observations shows the continuing importance of visual observations of Mars in reaching conclusions about Martian vegetation. Physical methods alone sufficed to prove the *possibility* of vegetation, but visual observations went a long way toward confirming that this possibility might actually be realized. Conversely, physical methods now confirmed that conditions gave credence to the vegetation hypothesis. No such conclusion was reached about cloud-shrouded Venus, which Coblentz and Lampland had determined to have a temperature of 50°C. Unlike Mars, no wave of darkening gave credence to Venusian plant life.

Coblentz's observations formed a turning point in Martian studies. For the first time, temperatures were determined by experiment to rise well above freezing on Mars. What is more, for the first time the dark areas were found to have a higher temperature than the light areas. These observed facts led Coblentz to an "assumption of the existence of plant life, in the form of tussocks, whether grass or moss," a form of vegetation having properties that could account for the higher temperature of the dark areas. But "beyond this assumption the writer withholds speculation," Coblentz declared.

Pathbreaking as these thermocouple observations were, the possibility of life on Mars depended on much more than the surface temperature of the planet; as Coblentz had noted, atmospheric pressure and composition would enable the possibilities of plant life to be further pinned down. Here a measure of the amount of planetary radiation, whether from a part of the planet or its integrated whole, was not enough. This was a problem for spectroscopy, and 14 years after Campbell's expedition to Mt. Whitney, the 60-inch reflector at Mt. Wilson was pressed into service for just this purpose, using the same technique with even higher dispersion. Using a six-prism spectrograph on February 2, 1925, Walter S. Adams and Charles E. St. John found that the water vapor in the atmosphere of Mars was 6 percent of that over the observing site at Mt. Wilson, indicating "extreme desert conditions over the greater portion of the Martian hemisphere toward us at the time." Similar measurements gave oxygen on Mars at 16 percent that above the Mt. Wilson site. These two independent measurements were the beginning of a series of downward revisions in the values of both Martian water vapor and oxygen. Like Petit and Nicholson, Adams and St. John drew no conclusions about Martian life from their work in their technical publications. But Adams was one of those in the *New York Times* symposium of 1928 who still lent support to the possibility of vegetation and even low animal life.

By the end of the 1920s, then, both visual observation and physical experiments were at one in strengthening the vegetation hypothesis. Visual methods still provided the best evidence, now supported by higher temperature measurements and less so by indications of very small amounts of oxygen and water vapor. Some of the participants, notably Trumpler, Coblentz, and Russell, were quick to discuss the implications of these results for life; others, like Petit, Nicholson, and Lampland, were not.

This fragile optimism of the 1920s was, however, soon jeopardized, most notably in 1933 by the renewed search of the Mt. Wilson director, Walter S. Adams, now in conjunction with Theodore Dunham, for oxygen on Mars. Using new photographic emulsions, a grating spectrograph, and the 100-inch telescope at Mt. Wilson, Adams and Dunham detected no oxygen and surmised an upper limit for Martian oxygen abundance less than 1/10th of 1 percent that of Earth. Though this result was discouraging for the hypothesis of Martian life, it did little to change belief in Martian vegetation during this period. Reviewing knowledge of the planets in 1938, Dunham himself wrote that, despite the lack of spectroscopic evidence of water or oxygen on Mars, the visual observation of a thin sheet of ice covering a small surface area was a more sensitive test than the spectroscopic attempts to find water spread over the entire atmosphere. And, as Russell had suggested, oxygen may have existed in the past and been locked in rocks to form ferric compounds, giving Mars its rusty color. Dunham's conclusion: "It would therefore be unwise to say that there may not be enough of both water vapor and oxygen to support life in some form which may have become gradually adapted to the rigorous conditions existing on Mars." This gradual adaptation by evolution was precisely the opinion of Russell himself after the observations of Adams and Dunham.

That the hypothesis of Martian vegetation survived, damaged but intact, between the late 1920s and 1945 is evident in the 1926 and 1945 editions of the standard textbook of the time, Russell, Dugan, and Stewart's *Astronomy*. Whereas Russell had written in 1926 that recent observations make it "very probable" that conditions on Mars support vegetable life, in 1945 he replaced those words with "not impossible," concluding in both editions that "it seems more likely than not that it does." But spectroscopic observations had taken a greater toll on the belief in animal life. Whereas in 1926 animal life had been viewed as "not impossible, or indeed, even improbable," by 1945 Russell wrote that "the great scarcity of oxygen makes the existence of animal life appear improbable." That the situation was in flux in both years Russell indicated by stating in both editions, "Recent evidence has greatly changed our estimate of the situation, and further evidence may change it again."

The wisdom of Russell's statement is evidenced in events following World War II, due in part to advances in technology made during the war for wartime purposes, in particular the lead sulfide cell sensitive to infrared wavelengths. Undoubtedly the premier event of the decade in Martian studies was the discovery by Gerard P. Kuiper (1905–1973) in 1947 of carbon dioxide (Fig. 2.9) in the atmosphere of Mars, a discovery of biological significance because it is a principal gas in the process of photosynthesis. The bright desert regions of Mars, Kuiper believed, were composed of igneous rock and the polar caps of water frost. As for the nature of the green areas, by comparing the Martian spectra with laboratory spectra of various plants on Earth, Kuiper found no evidence of chlorophyll characteristic of higher plants on Earth, and no evidence of plants containing water. Although the spectrum was consistent with the reflection spectrum of lichens, Kuiper declared in *The Atmospheres of the Earth and Planets* (1949) that this was by no means proof of lichens. In the second edition of 1952, however, Kuiper noted that "the minimum conditions for photosynthesis appear to be fulfilled," and he built a case for the possibility of lichenlike Martian plants, which would not show chlorophyll. A milder and more humid climate may have prevailed earlier, he argued, enabling slow adaptation, and volcanic vents may have played a role as a heat source. Although Kuiper remained cautious, urging that comparison with lichens might have only heuristic value and that final judgment be withheld, by the early 1950s the hypothesis of Martian vegetation had once again gained considerable ground. Similar conclusions were reached outside the Western world, based largely on the lifetime work of the Soviet astronomer Gavriil Adrianovich Tikhov (1875–1960).

As the favorable oppositions of 1954 and 1956 approached, interest in Mars, driven by interest in the Martian vegetation hypothesis, was increasing. As the 1956 opposition approached, Harvard astronomer William Sinton planned a direct search for vegetation by spectroscopic methods. Keenly aware that previous tests for infrared reflectivity characteristic of plants had been negative, Sinton's own search had a new element: it depended on the fact that organic

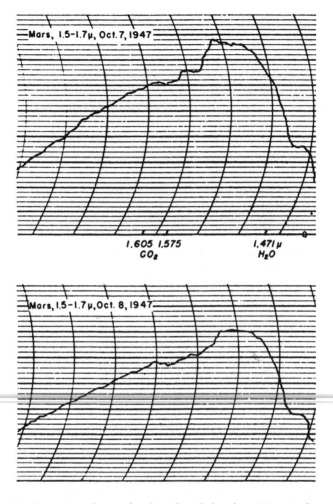

Fig. 2.9. G. P. Kuiper's evidence of carbon dioxide bands on Mars in the 1.6-micron region of the infrared spectrum, an observation confirmed by later evidence. Carbon dioxide is indicated by the two small dips on each of the plots, which were taken on 2 successive days. The observations were made with the infrared spectrometer on the 82-inch telescope at McDonald Observatory in Texas. From Kuiper's *The Atmospheres of the Earth and Planets* (Chicago, 1952), 360, by permission of The University of Chicago Press.

molecules have absorption bands at about 3.4 microns in the infrared part of the spectrum – beyond Kuiper's work that had been done in the 1- to 2.5-micron region. Sinton still used a lead sulfide photoconductive cell, as had Kuiper, but now cooled to 96°K with liquid nitrogen to increase its sensitivity to 3.6 microns. The difficulties of the observations can be appreciated from the fact that the sensitive area of this cell was only .16 mm square, and the diameter of Mars was less than 1 mm. Nevertheless, after 4 nights of observations, Sinton had enough evidence for his conclusion that the probability was "very high that an organic spectrum is required to account for the data."

Sinton was very much aware of previous visual evidence for vegetation in the form of seasonal changes in the size of the Martian dark areas. In fact, he saw the dip at 3.4 microns as "additional evidence for vegetation" and concluded in *Science* that "this evidence, together with the strong evidence given by the seasonal changes, makes it seem extremely likely that plant life exists on Mars." Thus, his graph of infrared absorption did not constitute direct visual confirmation but depended on the interpretation of spectrograms, an interpretation undoubtedly affected by preconceived ideas. The result caused considerable excitement, especially when Sinton confirmed it with equipment 10 times more sensitive on the 200-inch Palomar telescope during the 1958 opposition. Although the image of Mars was only 2 mm, this time Sinton separated the dark areas from the bright areas on Mars and confirmed his previous conclusion of absorption bands near 3.5 microns (Fig. 2.10). Again he concluded that "the observed spectrum fits very closely . . . that of organic compounds and particularly that of plants." In addition, a 3.67-micron absorption band was confirmed, which Sinton attributed to carbohydrate molecules in plants, analogous to tests on plants on Earth. Although Sinton's results were widely hailed and cited in the literature, they were open to interpretation: not only were other biological interpretations possible, by 1963 D. G. Rea, T. Belsky, and Melvin Calvin had done extensive work on infrared reflection spectra of terrestrial compounds and were critical of Sinton's interpretation. And by 1965 Rea, O'Leary, and Sinton himself suggested that two of the Sinton bands were due to heavy "deuterated" water (HDO) in the Earth's atmosphere, with the remaining band still possibly organic. In the end, the results of Sinton's refined methods had been rendered erroneous by refined problems. Just as V. M. Slipher a half century before believed he had found oxygen and water vapor on Mars, only to find that his results were contaminated by the Earth's own atmosphere, Sinton's results too were contaminated, this time by heavy water.

As with the canals of Mars, the search for Martian vegetation demonstrates differences in approach and worldview among scientists, with one extreme group much more likely to go out on a limb and to extrapolate than the other. Some astronomers probing the physical conditions on Mars presented their data and left them at that. Others used their data – indeed, were

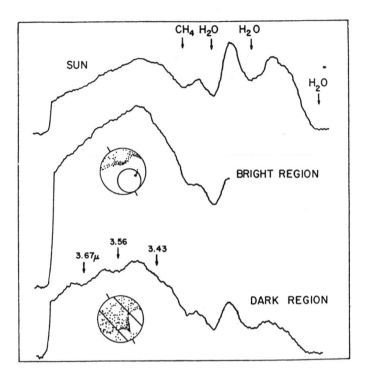

Fig. 2.10. Sinton's evidence of Martian vegetation. The top curve shows the solar spectrum, with absorptions by methane and water in the Earth's atmosphere. The middle curve shows a bright desert area of Mars where no vegetation was expected. The bottom curve shows absorptions from one of the dark areas of Mars, interpreted as due to vegetation. Unlike the data shown in Figure 2.9, this evidence turned out to be spurious in terms of revealing real characteristics of the planet. With permission from Sinton, "Further Evidence of Vegetation on Mars," *Science* (November 6, 1959), 1234. Copyright 1959 American Association for the Advancement of Science.

probably first inspired to gather their data – in the service of the question of extraterrestrial life. Still others rendered no opinion at all. In his book *Physics of the Planet Mars* (1954), Gerard de Vaucouleurs rarely mentioned the problem of life because "It is our belief that such a problem is still, to a large extent, beyond the limits of our positive knowledge and can only be the subject – either way – of vague speculations in which general 'principles' of a metaphysical nature have always to be taken as a guide."

To many, such a cautious attitude was not satisfying. They undoubtedly realized that the stakes in the debate extended far beyond Mars: as Kuiper wrote, "If life truly exists on the only two planets of the solar system that are at all suitable to sustain it, it is tempting to conclude that, after enough time has elapsed, it will develop spontaneously wherever conditions permit. Since planetary systems are presumed to be very numerous, life would then be no exception in the universe."

At the dawn of the Space Age, then, the canal controversy had receded, and much was known about the physical conditions of the planet Mars. Vegetation of some sort was still a very real possibility, dependent to some extent on what one saw as the limits to the adaptability of life. Vegetation did not have the popular appeal of intelligence, but to the scientist it was still a holy grail that held the promise of revealing the secrets of life. That promise was to play no small role in making Mars an important target for interplanetary probes of the Space Age.

2.3 THE SPACE AGE: LOWELL'S LEGACY OVERTURNED

By opening the possibility of space travel to Mars, the Space Age altered traditional approaches to the search for life in the solar system and finally brought consensus (though not unanimity) among scientists that extant life did not exist on the surface of the planet. The search for life via spacecraft required fundamental questions to be asked about the nature of life, for the first time bringing substantial numbers of biologists into an enterprise that had long been the province of astronomers. Although controversy marked the risky decision to search for life on Mars via spacecraft, once adopted that decision was a driving factor in the American space program. And though consensus on the results was eventually reached, observation via spacecraft – even when undertaken from the surface of the planet – still had its own problems.

Despite its novel methods, the beginning of the Space Age did not imply a rejection of all that went before or a diminution of the importance of ground-based studies. To the contrary, planning for expensive spacecraft brought a voracious appetite for whatever Martian data could be obtained by whatever means. Canals aside, Lowell's legacy continued to play a significant role, both as a driving force for exploration of the red planet and as a source of information, a fact that unifies the 20th-century search for life on Mars. Lowell left at least two legacies in his studies of Mars, one thoroughly discredited, the other still very much alive as the Space Age began. The discredited legacy was a planet crisscrossed by canals, built by inhabitants intent on making the best use of the remaining water on their arid planet. The lingering legacy was that of an arid but fundamentally Earth-like and possibly vegetated planet, with an atmospheric pressure of 85 millibars and caps of water ice that waxed and waned with the seasons. The persistence into the Space Age of Lowell's second legacy, and its final demise, are dramatic elements in a story never short on drama.

Even before planetary spacecraft became an imminent possibility with the launch of Sputnik in 1957, the effect of Lowell's second legacy is clear in early space biology discussions emerging from the discipline of space medicine. In his book *The Red and Green Planet: A Physiological Study of the Possibility of Life on Mars* (1953), for example, aviation and space medicine pioneer Hubertus Strughold used the Lowellian "green planet" as the backdrop to his discussion of "planetary ecology" and "ecospheres." Strughold agreed with Kuiper that Mars would most likely harbor only primitive plant life similar to lichens, and he suggested that the idea might be tested under simulated planetary conditions in the low-pressure chambers and climatic chambers used by aviation medicine over the last 20 years. The idea of simulated planetary conditions would prove a fertile one, and such experiments became a major activity carried out at many laboratories.

Meanwhile, the actuality of spaceflight and the imminent possibility of sending probes to the planets had set in motion another chain of events rising out of the concern of the larger biological science community, some of whose members realized the exciting possibilities now within their grasp but also foresaw that the very spacecraft designed to search for life might unwittingly destroy those possibilities. The Lowellian preoccupation with life – though now applied to microorganisms and much harsher conditions than Lowell had envisioned – was still at the heart of an idea that became a major concern of the space program: biological contamination. The contamination of the planets by some hardy terrestrial microorganism clinging to a spacecraft, or conversely and even more disastrously, "back-contamination" of the Earth by any returning samples or astronauts, was a prospect that had to be seriously considered, even in the context of such a seemingly hostile environment as the Moon.

As early as 1957, Joshua Lederberg (Fig. 2.11), soon to receive the Nobel Prize for his work in genetics, brought the issue of contamination before the scientific community. At Lederberg's urging, the following year the National Academy of Sciences of the United States urged the International Council of Scientific Unions (ICSU) to evaluate the problem and, if necessary, develop recommendations to prevent contamination. This they did through a Committee on Contamination by Extraterrestrial Exploration, which developed standards adopted by the ICSU in October 1958. But this was only the beginning. The contamination problem, as well as the problems of planetary life detection via spacecraft, was an important part of the discussion of two National Academy of Sciences Panels on Extraterrestrial Life, one on the East Coast (EASTEX) and one on the West Coast (WESTEX), which met from 1958 to 1960. Their efforts emphasized the seriousness of the problem. The ICSU's Council on Space Research (COSPAR) also addressed the issue.

Fig. 2.11. Joshua Lederberg, pioneer in exobiology, shown about 1962 in his laboratory at Stanford University.

The concern about contamination seemed to some hard-nosed engineers an obsession of wild biologists, but in the end it was taken very seriously. The contamination question was a contentious one that translated into real issues of economics, for spacecraft sterilization and astronaut quarantine were expensive programs. By the mid-1960s an extensive literature had developed on the subject, and one need only recall the quarantine of the returning astronauts and their samples to appreciate the attention given to back-contamination of the Earth. In order to shed light both on issues of contamination and on the conditions under which life might survive on another planet, scientists spared no efforts in studying the question of the adaptability of life in hostile environments on Earth.

Such programs to avoid biological contamination of the planets show how seriously the prospect of extraterrestrial life was taken. But nowhere was the Lowellian view of Mars more evident than in the exciting prospect that spacecraft provided the means for in situ examination of the planets for life. The biological interest in Mars was not by any means confined to a small group of scientists expanding the boundaries of space medicine or to those worried

55

about planetary contamination. Practitioners of the wider biological sciences were quick to realize the opportunities brought by the space program. In his influential paper "Exobiology: Experimental Approaches to Life Beyond the Earth," published in 1960, Lederberg was among the first to emphasize that space flight furnished a unique method for studying the most fundamental problems of biology, including the origin of life. To the study of the origins and evolution of life beyond our own planet he gave the name "exobiology."

NASA itself was quick to embrace exobiology as an important goal. In July 1959 NASA's first administrator, T. Keith Glennan, appointed a Bioscience Advisory Committee, which reported in January 1960 that NASA should not only be involved in a traditional and obviously necessary space medicine role in support of manned spaceflight, but that it should also undertake "investigations of the effects of extraterrestrial environments on living organisms including the search for extraterrestrial life." The committee, whose chairman and members represented various fields of the biological sciences, did not fail to see the same opportunities that Lederberg had earlier. "For the first time in history," they wrote with reference to the search for life and its origins, "partial answers to these questions are within reach." In the spring of 1960 NASA set up an Office of Life Sciences; by August, with the possibility of planetary missions on the horizon, it had authorized the Jet Propulsion Laboratory (JPL) to study the type of spacecraft needed to land on Mars and search for life. In order to study chemical evolution, the conditions under which life might survive, and a variety of related issues, NASA's first life sciences lab was also set up at its Ames Research Center in 1960.

In the Soviet Union, by comparison, the Russians had shown an even more focused interest in Mars as part of the larger "space race." The Soviets in fact had attempted unsuccessfully to send their first probe to Mars in 1960 and would persist in their attempts in the coming decades. Because the Cold War was in full swing, U.S. scientists had little knowledge of any details of the Russian Mars probes. They believed at the time that Soviet goals were similar to those in the United States, "with particular attention to the possibility of life." For reasons still unclear, however, the Russians seem never to have attempted any life detection experiments of the complexity contemplated by the Americans. Years later, it was reported that considerable effort was put into the idea of detecting the Sinton bands during the earliest flybys, but the Russian Mars landers carried only cameras and instruments to detect the chemical composition of the atmosphere and surface soil.

By contrast, interest in the United States was not only real, it was followed up by considerable action. Support for NASA's involvement received strong endorsement from the prestigious National Academy Space Science Board Summer Study at Iowa State University in 1962, which reviewed the full scope of space biology. In setting forth the general philosophy for the three areas of space biology – exobiology, environmental biology, and man in space – the authors were unequivocal in "setting the search for extraterrestrial life as the prime goal of space biology." In doing so, they realized not only the scientific but also the philosophical import of exobiology: "it is not since Darwin – and before him Copernicus – that science has had the opportunity for so great an impact on man's understanding of man. The scientific question at stake in exobiology is, in the opinion of many, the most exciting, challenging, and profound issue, not only of this century but of the whole naturalistic movement that has characterized the history of western thought for three hundred years. What is at stake is the chance to gain a new perspective on man's place in nature, a new level of discussion on the meaning and nature of life." With this statement and the detailed recommendations that followed, exobiology was firmly entrenched as a major NASA effort and – not coincidentally – one with high public interest. Two years later, the Space Science Board reaffirmed in its report *Biology and the Exploration of Mars* that "the exploration of Mars – motivated by biological questions – does indeed merit the highest scientific priority in the nation's space program over the next decade."

With a strong consensus on carrying out a search for life on Mars via spacecraft, the next question was exactly how to implement this consensus. In March 1959, the University of Rochester biologist Wolf Vishniac had received a NASA grant to develop "a prototype instrument for the remote detection of microorganisms on other planets." Proposals had proliferated so much by 1963 that NASA's Ames Research Center set up a team to evaluate the concepts. Photographic reconnaissance spacecraft to Mars were in the design stage in the early 1960s, and spacecraft with real experimental capability in the realm of life detection were the ultimate goal.

In short, for the first time, substantial government resources were being invested in the search for extraterrestrial life. This was not Percival Lowell and his small staff chasing an eccentric theory or even a scattering of astronomers around the country at their observatories trying to prove or disprove a theory. This was nothing less than a national government proposing to spend substantial sums of taxpayer money to launch a concerted effort to search for life beyond the Earth. For the first time, therefore, the question of extraterrestrial life entered the arena of science policy. It is not surprising in a democratic society that this NASA goal – like many others – was challenged and criticized. Among the critics were two prominent scientists: chemist Philip H. Abelson and microbiologist Barry Commoner. Both questioned the goal not only for scientific reasons, but also as a matter of priorities in a world filled with poverty. Abelson used his position as editor of the influential journal *Science* to argue in 1965 that "In looking for life on Mars we could establish for ourselves the reputation of being the greatest Simple Simons of all time."

Nevertheless, spacecraft exploration of Mars enjoyed widespread public support in the United States and had developed a tremendous momentum that could not be halted even when new ground-based studies and the earliest spacecraft yielded surprisingly discouraging results (Table 2.3). We recall that by 1963 Sinton's claim for Martian vegetation had been brought into question, and that by 1965 Sinton himself believed the infrared bands were not due to Martian vegetation at all but rather to deuterated water in the Earth's own atmosphere. In 1963 too, Spinrad, Münch, and Kaplan reported from spectroscopic observations at Mt. Wilson small amounts of water vapor and an extremely low Martian atmospheric pressure of 25 millibars. And when in July 1965 Mariner IV passed within 6,118 miles of Mars and relayed 22 photos showing a cratered and apparently dead planet, its instruments measured an even lower atmospheric pressure of 10 millibars at the surface. This was only 1 percent of the Earth's 1,000-millibar surface atmosphere, the equivalent of Earth's atmosphere at 90,000 feet. Though all these developments supported Abelson and other critics, the National Academy of Sciences Board, in a Postscript to *Biology and the Exploration of Mars,* did not alter the recommendations of its final report, and NASA pressed ahead. By early 1969 Mariners 6 and 7 had taken high-resolution photographs of about 20 percent of the Martian surface, with results that did nothing to improve the bleak outlook for life on Mars.

But as if to underscore the need for caution in reaching preliminary conclusions, the mission of Mariner 9 in 1971 showed a very different Mars – one with many kinds of terrain, including channels that resembled nothing so much as dry river beds. If there had once been water on Mars, might not life have existed – and perhaps survived? Mariner 9 – a milestone as the first spacecraft to orbit another planet, returning numerous images of high resolution – rejuvenated such questions, but at the same time it permanently laid to rest the Mars of Percival Lowell.

The culmination of the search for life in the solar system was the landing of two Viking spacecraft on the surface of Mars in 1976, surely one of the great adventures in the history of science and technology. The Viking project, initiated in 1968 after the demise of the Mars Voyager project and now managed by NASA's Langley Research Center, was an example of "big science" at its best in terms of budget, staff, goals, and results. The cost of the Viking spacecraft, including the orbiters, landers, and support (but not launch vehicles), was $930 million. Although the usual funding hurdles had to be overcome and many critics answered, in the end two Viking orbiters arrived at the planet on June 19 and August 7, 1976. After suitable reconnaissance, as the United States celebrated its bicentennial back on Earth, two Viking landers set down on Mars in July and September. Under the guidance of project scientist Gerald A. Soffen, 13 teams with a total of 78 scientists undertook 13

Table 2.3. *Space Age observations of Mars relevant to life*

Observer	Date	Location	Method	Result
Sinton	1959	Palomar 200 in.	IR spectroscopy	Vegetation
Spinrad/ Münch/ Kaplan	1963	Mount Wilson 100 in.	IR spectroscopy	Water vapor 25-mb pressure
Mariner 4	Nov. 28, 1964	Mars flyby	Photography	Impact craters
Mariner 6	Feb. 24, 1969	Mars flyby	Photography	20% of surface photographed
Mariner 7	Feb. 18, 1969	Mars flyby	Photography	
Mariner 9	May 30, 1971	Mars orbit	Photography	Entire surface photographed; channels
Viking 1 lander	July 20, 1976 (landed)	Mars surface (Chryse Planitia)		
Viking 2 lander	Sept. 3, 1976 (landed)	Mars surface (Utopia Planitia)		
Viking biology experiments				
Oyama/ Berdahl			Gas exchange	O_2 and CO_2 liberated, no life
Levin/ Straat			Labeled release	Positive result but ambiguous
Horowitz/ Hobby/ Hubbard			Pyrolytic release (carbon assimilation)	Positive result but not biological
Biemann, Oró, et al.			Gas chromatograph mass spectrometer (GCMS)	No organic compounds
Hess et al.			Meteorological	Pressure 7 mb average at surface; temperature 180–240 K over 1 day
Meteorite fall				
McKay et al.	Aug. 1996	Antarctica Allan Hills "Mars rock"	Meteorite analysis	Organic molecules, carbonate globules, magnetic minerals, microfossils?

separate investigations, including 3 mapping experiments from the orbiter, 1 atmospheric experiment, 1 radio and radar experiment, and 8 surface experiments. The total cost for development and execution of these experiments was another $227 million. The results increased knowledge of Mars far beyond that of all previous investigations combined, finally providing definitive answers to age-old questions, including the issues of temperature, atmospheric composition, and pressure so crucial to life.

From beginning to end, though the various science teams grappled with the myriad problems of meteorology, seismology, chemistry, imaging, and physical properties of the planet Mars, the Viking biology experiments were the driving force behind the project, as evidenced by both budget and public, congressional and even scientific interest. A total of $59 million was spent on the Viking biology package and another $41 million on the molecular analysis experiment that was relevant to the question of life because of its ability to detect organic molecules. Harold P. Klein of NASA's Ames Research Center headed the Viking biology science team; Klaus Biemann of MIT headed the separate molecular analysis team. While the results of several of the teams were relevant to the question of Martian biology, these 2 out of the 13 teams were most directly relevant.

The Viking biology package (Fig. 2.12) embodied in one piece of technology the most sophisticated thinking of the 20th century on the subject of extraterrestrial life in the solar system. The assumptions behind its experiments, the results obtained, and the ensuing controversies over the interpretation of these results are therefore of considerable importance. The diverse ideas about the nature of Martian life led to three different biology experiments aboard Viking, each representing a different approach to the problem of life. Indeed, biology team leader Klein later stated that had it not been for the constraints of 15-kg weight and about 1 cubic foot volume for the biology package, even more of the approaches conceived during the previous two decades would have been included on the spacecraft. The idea was that the three experiments (Table 2.3), singled out and recommended by the Space Science Board of the National Academy of Sciences in 1968, would test for life using different philosophies, environmental conditions, and detectors.

One approach, which came to be known as the "labeled release" experiment, was developed by Gilbert Levin, who had spent much of the 1950s trying to improve methods for the detection of bacterial contaminants in city water supplies and believed his method could be applied to the search for life on Mars. He was awarded a NASA contract for his "Gulliver" concept in 1961 and was reporting on his experimental apparatus in the early 1960s. Levin's approach assumed that any Martian microorganisms, like those on Earth, would assimilate (eat) simple organic compounds, decompose them, and produce gases such as carbon dioxide, methane, or hydrogen as end

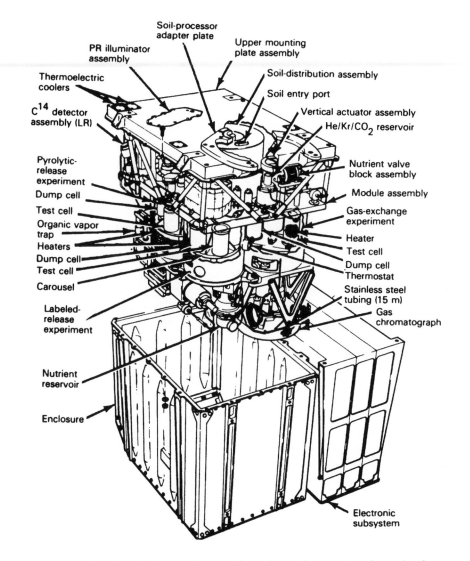

Fig. 2.12. The Viking biology package, with a volume of approximately 1 cubic foot, showing the three biology experiments and associated equipment. The pyrolytic release experiment is at the upper left, the labeled release experiment is at the lower left, and the gas exchange experiment is at the center right. Courtesy NASA.

products. For this reason, a dilute aqueous solution of seven such organic compounds, radioactively labeled for detection purposes, was added to the incubation chamber containing the Mars soil sample. The experiment tested for the expected labeled release of the gas produced as any organisms ate the organics and breathed out the decomposition products. The output was in the form of radioactive disintegrations, measured by a carbon-14 detector in counts per minute.

The second biology test, the "gas exchange" experiment, was developed by Vance Oyama of NASA Ames Research Center, a veteran of life detection experiments on Apollo lunar samples. The gas exchange experiment tested for life under two different conditions. In the first mode, it was assumed that any organism in the dry Martian environment would be stimulated to metabolic activity by the addition of slight water moisture, and would give off a gas that could be detected by chromatography in the area immediately above the sample. In the second "wet nutrient" (or "chicken soup") mode, a rich nutrient of 19 organic compounds was added as an additional stimulus to metabolic activity, the products to be detected in the same manner. In both cases, the liquid added did not come into contact with the soil but was added underneath the cell in which the soil "incubated." Water vapor gradually seeped up through the porous bottom of the incubation chamber, creating gradations of moisture through the soil. Experiments were also undertaken without the addition of any moisture.

The "pyrolytic release" experiment (also called the "carbon assimilation" experiment) was headed by Norman Horowitz of Caltech. Horowitz, a member of the WESTEX group in 1959, had cooperated with Levin's project in the early 1960s, but after Mariner IV showed that liquid water could not exist on Mars, he split with Levin and became convinced that it was best to test for Martian organisms under conditions known to exist on Mars when the experiment was designed. Thus, to the small sample of Martian soil Horowitz proposed in his experiment to add only carbon dioxide and carbon monoxide, gases known to exist in the Martian atmosphere and now radioactively "tagged" for detection purposes. It was assumed that any organism on Mars would have developed the ability to assimilate these gases and convert them to organic matter. After 120 hours of incubation, the soil chamber was to be heated to 635°C to pyrolyze the organic matter and release the volatile organic products, hence the name "pyrolytic release." A radiation counter yielded disintegrations per minute.

All three experiments sought to detect metabolic activities. Of the experiments, Oyama's wet nutrient mode was the most Earth-like approach in that it added rich terrestrial organics to stimulate any Martian organisms. Horowitz's was the most Mars-like, making few assumptions about Martian life except that it would be carbon based. Levin's, with his weak organic nutrient, fell in between. Levin and Oyama attempted to detect life by the decomposition of organics into gas during metabolism (a universal property of terrestrial organisms), while Horowitz sought to synthesize organic matter, which he would then pyrolyze in order to be able to detect it. For detection purposes, both Levin and Horowitz made use of standard techniques of radioactive carbon-14 as a "tracer," a method that did not change the chemistry but provided a means of distinguishing atmospheric carbon from metabolized carbon. Oyama made use of the well-known method of gas chromatography for detection, as did Biemann (in conjunction with a mass spectrometer) for the organics experiment, which had nothing to do with metabolism. Since they were ignorant of the nature of Martian life, the fondest hope of all the experimenters was that at least one of the experiments – hopefully their own – would turn up something.

Summer 1976 finally brought the day that Lowell, Kuiper, and a host of scientific ghosts would have savored: the landing of two spacecraft on the surface of Mars to test for life in situ. They would not have been disappointed: Viking 1 landed successfully on the Chryse plain on July 20, and the first results of the biology experiments returned from Viking were exciting, to say the least. Although no visible life forms walked across the field of view of the camera, once the soil samples were collected on July 28, the biology experiments quickly began to return major surprises. Levin's experiment evolved gas into the chamber after the nutrient was added; then the reaction tapered off. Horowitz's pyrolytic release test was also positive, and Oyama's gas exchange experiment evolved not only carbon dioxide but also oxygen, the latter a reaction never before seen in tests on terrestrial or lunar soils. Because of the speed and course of the latter reaction, Oyama's experiment was not believed to be biological in nature. In short, two of the three biology experiments gave "presumptive positive results" for biology, and the third gave evidence of an oxidizing material in the surface at the Viking site. There was only one problem: in another unexpected finding, Biemann's organic analysis showed no organic molecules present to the level of a few parts per billion, a result later called "probably the most surprising single discovery of the mission." As Klein subsequently recounted, these first results caused the carefully laid-out experimental strategy to be abandoned as the scientists attempted to discover whether chemical or biochemical reactions were taking place.

By 8½ months after Viking 1 had landed, 26 biological experiments had been carried out and the first relatively complete results were reported, along with other Viking experiments, in the *Journal of Geophysical Research*. By then, shortly before the biological experiments were terminated in May 1977, Klein's considered judgment was that the positive result of Horowitz's pyrolytic release experiment was probably nonbiological in origin, while Levin's labeled release experiment remained ambiguous. Ironically, the gas exchange experiment of Oyama – the scientist most optimistic about Martian life after Vishniac's death – showed no evidence at all of biological activity. Oyama and most of his colleagues concluded that the spontaneous evolution of oxygen was due to a chemical reaction involving "superoxides" such as hydrogen peroxide, perhaps by the effect of solar radiation on the small amount of water vapor in the upper atmosphere of Mars. "It's like the three bears," Klein later said. "Not too much water, not too little water, just the right amount of water in its atmosphere to produce something like this. This is one of the big mysteries, and any future missions to Mars have to find out what this stuff is."

In the end, there was no complete consensus among the experimenters themselves. Writing for *Scientific American*, Horowitz concluded that although "it is not easy to point to a nonbiological explanation for the positive results" of his pyrolytic release experiment, "it appears that the findings of the pyrolytic-release experiment must also be interpreted nonbiologically," mainly because the reaction was less sensitive to heat than one expected from a biological process. Levin, however, did not agree; in the 1980s he continued to argue forcefully that a biological interpretation of his data was still possible.

Clearly sensitive to their own assumptions, the Viking biologists continued to ponder the strategy of their experiments. What if their assumptions about Martian life, on which the biology experiments were based, were not correct? With this in mind, Klein concluded his summary of Viking biology results with the astonishing remark that "we must not overlook the fact, in assessing the probabilities of life on Mars, that all of our experiments were conducted under conditions that deviated to varying extents from ambient Martian conditions, and while we have accumulated data, these and their underlying mechanisms may all be coincidental and not directly relevant to the issue of life on that planet."

After 10 years of contemplation of the experiments conducted for some 10 months on the surface of Mars, Horowitz remained convinced that they not only proved the absence of life on Mars, but by extension, "Since Mars offered by far the most promising habitat for extraterrestrial life in the solar system, it is now virtually certain that the earth is the only life-bearing planet in our region of the galaxy." Although most scientists were not ready to make that quantum leap, it is also fair to say that they were much less optimistic about life on Mars in the aftermath of Viking. The Viking results were impressive enough that most scientists shifted the focus of their biological Martian interests to either past Martian history or different Martian environments such as rocks, polar caps, subsurface soil, or volcanic regions. Certainly the ancient river valleys still remained a major challenge, and it would have been one of the striking ironies of history if, a century after Lowell's canal hypothesis, the "channels of Mars" gave evidence of fossil life. Twenty years after Viking, a startling announcement of evidence of Martian fossil life opened just that possibility and brought the Martian life controversy back to the forefront.

2.4 REPRISE: MARTIAN FOSSILS AND EUROPAN SEAS

As the 20th century approached its end, it appeared that the Viking landers had written the last chapter in the search for life on Mars. But in August 1996, the world was startled with the announcement that organic molecules, possibly biogenic minerals, and even microfossils may have been found in a meteorite that originated on Mars. The result was controversial, but the inconclusive evidence was more than balanced by the fact that the Martian meteorites could now be examined, not with the limited resources of a spacecraft on the surface of Mars, but with the full power of analytical techniques in many laboratories on Earth. A new era in Martian life studies had begun.

Meteorites had long been associated with the question of extraterrestrial life and the origin of life, as we will see in Chapter 6. But those meteorites were a special variety known as "carbonaceous chondrites," and their parent body had not been identified. Only in the post-Viking era was a new category of extremely rare meteorites identified and a case slowly built that they had originated on Mars. Known as "SNC meteorites" after the locations of their three types (shergottites, nakhlites, and chassignites), they were also stony meteorites but were "achondrites" because they exhibited none of the millimeter-sized embedded mineral spheres characteristic of chondrites. They were known to have come from Mars not only because of their chemical composition, but also because the gases trapped in them had precisely the same composition and proportions as those of the Martian atmosphere, as determined by the Viking landers. Thus, although Viking did not unambiguously find life itself, ironically it enabled the identification of the SNC meteorites as Martian in origin.

The surprising announcement in the summer of 1996 centered on the Martian meteorite known as Allan Hills 84001 (Fig. 2.13), believed to have fallen on the ice fields of the Antarctic 13,000 years ago. The first meteorite found in the Antarctic during a National Science Foundation-sponsored search season in 1984 (hence the name ALH 84001), it was not identified as Martian in origin until 1994. One of only 12 such meteorites identified at the time, the 4.5-lb (1.9-kg) softball-sized rock was by far the oldest of the 12, estimated to have formed about 4.5 billion years ago, in a period when Mars was warmer and had water and an atmosphere. It was hypothesized that a meteorite impact on Mars fractured the rock about 3.6 billion years ago, and that another impact about 16 million years ago launched the rock into space, where it eventually intercepted the Earth.

Fig. 2.13. ALH 84001, the 4.5-billion-year-old Mars rock found in the Antarctic, believed by some to contain fossil evidence of primitive Martian life. The rock weighs 4.5 lb (1.9 kg), and the scale is in centimeters. Courtesy NASA.

The evidence, announced by a NASA team led by David McKay of NASA's Johnson Space Center in Houston, consisted of four parts. None of these parts, the participants pointed out, was conclusive in itself, but taken together they could be interpreted as biogenic. First, the multidisciplinary science team reported, the fractured surfaces of the rock contained large, complex organic compounds in the form of polycyclic aromatic hydrocarbons (PAHs). This was already a step beyond what the Viking landers had found, but even though the NASA team undertook analysis that showed to their satisfaction that the PAHs were not contamination from Earth, this was not proof of life since organic molecules could have originated by nonbiogenic processes on Mars. But then the plot thickened: in the fractures the team also discovered carbonates, mineral deposits that may be produced by living creatures on Earth, as in the case of limestone. Within the carbonates they also found magnetite, pyrrhotite, and greigite, minerals that are produced (among other ways) by certain "magnetotactic" bacteria on Earth. Finally, using a high-resolution scanning electron microscope, the team suggested the existence of

66

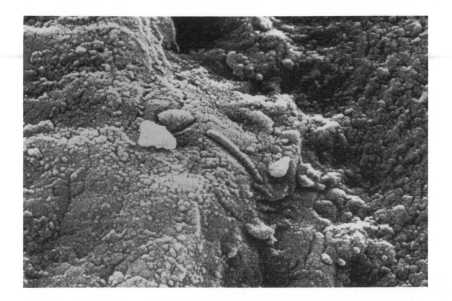

Fig. 2.14. High-resolution scanning electron microscopic image showing an unusual tubelike structure less than 1/100th the width of a human hair, found in Martian meteorite ALH 84001 and interpreted by some to be evidence of fossil life on Mars. Courtesy NASA.

microfossils in the carbonates and other mineral grains (Fig. 2.14); at only 20 to 100 nanometers they were 100 times smaller than the smallest known bacteria on Earth.

Less than 2 months later, a British team of scientists led by Colin Pillinger of the Open University announced independent evidence of possible traces of life, both in ALH 84001 and in a much younger Martian meteorite known as Elephant Moraine 79001 (EETA 79001, again named after the location of its discovery in the Antarctic). The latter meteorite was only 175 million years old and was blasted from Mars only 600,000 years ago. This was so recent, geologically speaking, that it held open the possibility that life might still exist on Mars.

As in past controversies over Martian life, the stakes were high and the skeptics numerous. One of the chief objections came from Ralph Harvey of Case Western and Harry Y. McSween of the University of Tennessee, who had reported in *Nature* shortly before the NASA announcement that their analysis of the same meteorite showed that the carbonates formed not as a result of microbial life, but during the asteroid impact when carbon dioxide combined with the rock at temperatures of 700°C. Such temperatures are inimical to life; if this method of carbonate formation was confirmed, it would cast severe doubt on the claims of past Martian life. Others, however, argued for low-temperature formation of the carbonates, one that did not rule out life. Definitive proof of past life on Mars would come only by sectioning thin sections of the microfossils to search for cell walls, DNA, or other structures unambiguously linked to life. In the years following the announcement, many teams were doing precisely that, even as other parts of the rock were sent to dozens of other laboratories for microscopic analysis.

These claims produced intense interest not only among the public, but also at the highest levels of government. In the United States, in December 1996, Vice President Al Gore gathered together some 20 experts from a wide variety of disciplines to discuss the implications if the discovery of life on Mars were true. NASA considered how plans for Mars exploration should be changed in the context of its "Origins" program; exobiology research was given increased funding even as a new niche called "astrobiology" sprang up. The British team made its announcement at a meeting of the Royal Society of London hosted by the British minister of science. Scientific journals and the popular press around the world echoed the excitement, tempered only by the still uncertain results. Mars was once again the planet of mystery, with life as a driving force.

Adding to the excitement over life in the solar system was the discovery by the American Galileo probe, also in August 1996, that the Jovian moon Europa might harbor a biosphere in "warm ice" or even in an underground liquid ocean. The evidence was in the form of photographs (Fig. 2.15) showing images resembling ice floes on Earth's polar regions and enormous cracks thousands of miles long in the ice that covers the surface of Europa, an object about the size of the Earth's moon. Originally discovered by the Voyager 2 spacecraft in 1979, the cracks are believed to be caused by tidal stressing from Jupiter's enormous gravity. The new images added impetus to the theory, also celebrated in science fiction, that Europa may harbor life in oceans that are warmer and deeper than was previously thought, perhaps 50 miles deep. The idea also gained increased support in light of findings that life on Earth flourishes deep underwater in complete darkness, drawing energy from planetary heat in the form of volcanic vents rather than from sunlight. Europan life, too, remained uncertain, and was certain to drive yet another research program encompassing what some were calling "exo-oceanography."

At the end of the 20th century, then, expectations for life in the solar system had been lowered but by no means extinguished. The extraordinary interest in planetary life sparked by Lowell not only continued but was surprisingly extended to other potential habitats in the solar system as more was learned about the incredible tenacity of life. While the importance of Lowell's legacy is evident even in the Space Age, what also stands out is the extraordinary

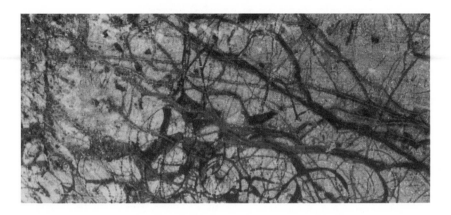

Fig. 2.15. Fractured surface of Jupiter's moon Europa, which indicates to many scientists that a liquid ocean may exist below the ice. Some areas resemble ice floes seen in Earth's polar seas. Europa is about the size of Earth's Moon; this image is about 220 by 475 miles (360 by 770 km), and the smallest visible feature is about 1 mile (1.6 km). Galileo spacecraft image taken June 27, 1996. Courtesy NASA/JPL.

progress made during the century on the question of life on Mars, despite Lowellian preconceptions. Grave difficulties and problems of interpretation notwithstanding, the process of observation eliminated from Mars first intelligence, then vegetation, and finally organic molecules. Yet, even if the claims of life in Martian meteorites proved to be spurious, the dry river channels on Mars were sure to spur speculation that life may have arisen on a warmer, wetter Mars in the past. And then there was Europa, and aside from Europa, Saturn's moon Titan, which with its organic molecules also remained on the list of potential exobiological sites. Thus, contrary to post-Viking expectations, it seems likely that Mars, Europa, and Titan will continue to tantalize exobiologists well into the 21st century, and that further exobiological surprises may await us even in the solar system.

3

SOLAR SYSTEMS BEYOND

We begin to suspect that life is not the normal accompaniment of a sun, since planets capable of sustaining life are not the normal accompaniments of suns. Astronomy does not know whether or not life is important in the scheme of nature, but she begins to whisper that it must necessarily be somewhat rare.

James Jeans (1923)

Millions of planetary systems must exist, and billions is the better word. Whatever the methods of origin, and doubtless more than one type of genesis has operated, planets may be the common heritage of all stars except those so situated that planetary materials would be swallowed up by greater masses or cast off through gravitational action.

Harlow Shapley (1958)

Should we not come to the rescue of a cosmic phenomenon trying to reveal itself in a sea of errors?

Peter van de Kamp (1983)

The search for extrasolar planets can be amazingly rich in surprises. From a complete planetary system detected around a pulsar, to the rather unexpected orbital parameters of 51 Peg b, searches begin to reveal the extraordinary diversity of possible planetary formation sites.

Michel Mayor and Didier Queloz (1995)

The widespread existence of life beyond the Earth, by any standard definition of life, required that planetary systems be a common feature of the universe. Although the existence of such systems became an increasingly important research problem in astronomy during the 20th century, the observation of planetary systems was extremely difficult. Whereas Mars had posed problems when observed only tens of millions of miles away, even the nearest stars were a million times more distant. The direct observation of planetary companions to those stars was out of the question for most of the century and still problematic at its end. Whereas in our solar system the crucial problem was to observe planetary surfaces and atmospheres, outside the solar system only the *effects* of supposed planets could hope to be observed, not the planets themselves. Even here the difficulties were so great that, as observation played the central role in the search for life in the solar system, theories of solar system formation assumed the dominant role for much of the century in the quest for other solar systems. The bearing these theories had on the idea of other planetary systems, though not always explicit, was never far below the surface. Sir Harold Spencer Jones stated it clearly in his *Life on Other Worlds* (1940): "if we can find out how the solar system came into being we shall possibly be able to judge what likelihood there is that other stars have families of planets." But theory could never consummate the search for planetary systems, and although as late as 1994 the search for planets around Sun-like stars remained elusive, by century's end observation finally came to the rescue after a long history of spurious claims.

3.1 SKEPTICISM: CLOSE ENCOUNTERS OF THE STELLAR KIND

At the turn of the 20th century scientific opinion on the subject of planetary systems was far from unanimous. The late 19th century had generally favored the notion of abundant planetary systems, often for philosophical reasons. On the other hand, as more and more multiple star systems were discovered, some argued that planetary systems were rare because conditions for life could be favorable only around single stars, and only certain types of single stars at that. The undecided scientific mind was therefore quite open to any evidence – theoretical or observational – that would shed light on the subject of planetary systems. In the absence of direct observation of planets, prohibited by their relatively small size and the glare of the stars they were supposed to surround, astronomers necessarily resorted to indirect methods.

The most common claim for the observed effects of planets came in the debate over "unseen companions" of stars, as revealed by the gravitational perturbation of a star by one or more encircling planets. Friedrich Wilhelm Bessel had measured variations in the motions of the stars Sirius and Procyon in 1844 and argued that these were due to the presence of unseen companions. Although these companions turned out to be stars rather than planets, with the detection of such "wobbles" astronomy had a method for planet detection if only delicate enough observations could be made. Aside from this "astrometric" method, two other planetary effects on a central star were possible. A change in the amount of light seen from the star (the photometric method) would be caused simply by the slight diminution of light as a planet passed in front of the star as seen from Earth, an argument that had its roots in the postulation in 1782 of an eclipsing companion of the variable star Algol. And a change in the spectrum of the star (the spectroscopic method) would be caused by changes in the radial velocity of the star as a planet tugged it one way or the other in the course of its orbit, an effect first detected for a companion star by E. C. Pickering with the discovery of "spectroscopic binaries" in 1889. These three techniques – the astrometric, the photometric, and the spectroscopic – were all proven methods for the detection of unseen stellar companions by the turn of the century and would still maintain their validity as planet detection techniques at the end of the century, albeit with a much greater appreciation of the difficulties involved when searching for objects of low mass. Despite numerous attempts using these three indirect observational methods, not until the very end of the century would they yield unambiguous results for planet-sized objects.

In the absence of decisive observational evidence for planetary systems, one might expect that theories of solar system formation would play an especially important role, at least in determining the plausibility of such systems. This had indeed been the case for the nebular hypothesis, which favored abundant planetary systems because the formation of planets from a rotating gaseous disk was assumed to be a universal process. But that hypothesis was under serious attack in 1900, and for the first two decades of the 20th century the new theory, to the limited extent that it addressed the issue at all, gave conflicting indications about the possibility of other planetary systems. Developed by T. C. Chamberlin, chairman of the Geology Department at the University of Chicago, and F. R. Moulton, an astronomy graduate student at the same university, the Chamberlin–Moulton hypothesis sought to surmount the technical weaknesses of the nebular hypothesis by proposing instead that solar systems were formed by the close encounter or actual collision of stars in space (Fig. 3.1). According to this hypothesis, the close encounter caused material to be ejected from the Sun. The passing intruder then caused the ejected material to form spiral arms. These arms contained knots of denser material that condensed into nuclei, which in turn grew into planets and satellites by the capture of "planetesimals," cold particles in the nebula. The spiral nebulae recently observed in the heavens, they believed, might be evidence of such collisions and of solar systems in formation. Curiously, rarity or abundance of planetary systems does not seem to have been an issue for Chamberlin or Moulton. To the extent that their rarity or abundance was an issue at all, it oscillated between the twin pillars of the "planetesimal hypothesis": the spiral nebulae, which implied abundance, and stellar encounters, which implied rarity. With the gradual realization that spiral nebulae were too large to represent planetary systems in formation, the stellar encounter aspect of the theory was free to gain the upper hand – and with it the implication of the rarity of planetary systems.

This, in fact, is precisely what occurred, not in America but in Britain, where in the tradition of Whewell and A. R. Wallace (though not of Richard Proctor), the scientific community seemed more skeptically inclined toward planets and life. It was at the hand of the British mathematical physicist and astronomer James Jeans that the question of other solar systems would become closely linked with the rarity of planets and life in the universe. Jeans, a 1903 graduate of Trinity College, Cambridge, had done important work on atomic theory and statistical mechanics prior to 1914. After 1914 he turned from the microscopic to the macroscopic, from atoms to astronomy, and specifically to cosmogony. Jeans's attention was at first devoted to the stability of rotating bodies, on which subject he published two lengthy papers in

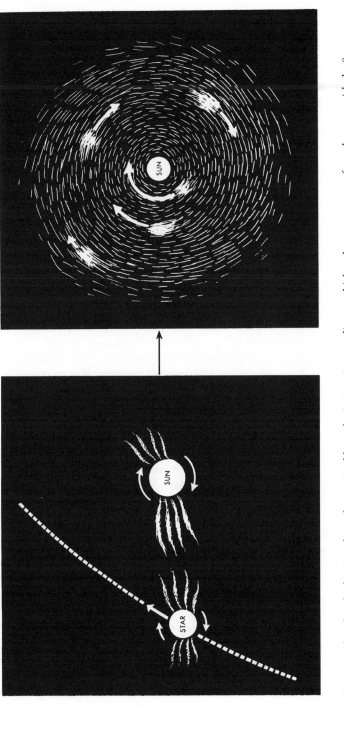

Fig. 3.1. The Chamberlin–Moulton planetesimal hypothesis (1905), according to which a close encounter of another star with the Sun causes gases to erupt from both (left). These gases condense to form a large number of planetesimals that, in turn, accrete to form the planets (right). The spiral part of the theory, dropped a few years after it was proposed, is not depicted. From a review of theories of the origin of the Earth by the astronomer Thornton Page, *Physics Today* (October 1948), at a time when abundant planetary systems were again being proposed.

1915 and 1916. This work he applied to cosmogony in 1916 with reference to tidally distorted masses – in other words, determining how a rotating astronomical body would be affected by tidal forces raised by another passing astronomical object, as would happen in the case of a close stellar encounter. In a paper read before the Royal Astronomical Society in 1916 and published in the society's *Memoirs* the following year, Jeans dealt not only with the origin of solar systems, but also with binary star formation and spiral nebulae. In contrast to the binaries and spirals, Jeans concluded that the solar system might well have been formed from a tidally distorted mass, in particular by another star approaching our sun. Unlike the Chamberlin–Moulton hypothesis, however, Jeans's analysis showed that neither spiral nebulae nor planetesimals played a role in planet formation, and he thus emphasized that for solar systems "the origin which seems most probable is not that of the planetesimal hypothesis." Instead, his analysis showed that rather than the streams of gas torn from the Sun condensing into numerous small, cold planetesimals that in turn accreted to form the planets, a single cigar-shaped filament of hot gas would be ejected and condense directly into the planets (Fig. 3.2). As the theory was later elaborated, he pointed out that the largest planets would form near the center, where the filament was thickest, and the smaller ones at each end, giving the distribution of planets observed in our solar system.

The central question in determining whether this mathematical conclusion could really occur in nature was the frequency of close stellar encounters. It is clear at the outset of the paper that Jeans was already thinking in these more general terms, not only with regard to the origin of our solar system, but in connection with the frequency of planetary systems. In his earliest statement on what would become a lifelong contentious issue, he wrote in his paper on tidally distorted masses that "We have absolutely no knowledge as to whether systems similar to our solar system are common in space or not. It is quite possible, for aught we know to the contrary, that our system may have been produced by events of such an exceptional nature that there are only a very few systems similar to ours in existence. It may even be that our system is something quite unique in the whole of space."

Jeans's analysis showed that the issue of abundance was very sensitive to the assumptions made about a variety of parameters, including the density of stars in the universe, the velocity of the stars in space, the age of the stars and of the universe, and the size and mass of the stars at the time of encounter. All of these parameters were subject to change in the discussion that ensued over the next three decades. For now, using the best estimates known in 1916 and assuming stellar masses and velocities similar to those of the Sun, Jeans found that at most 1 star in 4000 might have experienced a "nontransitory" encounter at the distance of Jupiter in a lifetime of 10 billion years, the upper limit that he placed on the age of the universe. If the encounter

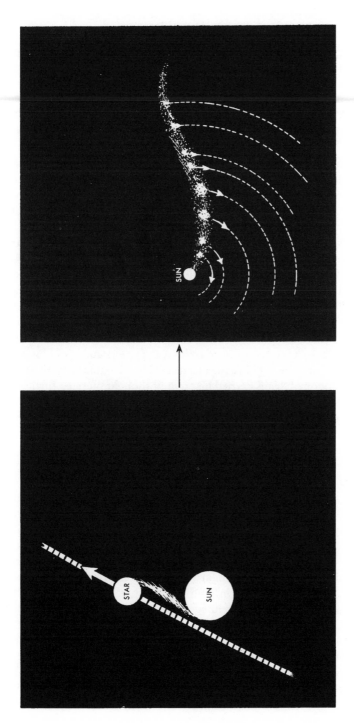

Fig. 3.2. The Jeans–Jeffreys hypothesis (1917) has the passing star pull out a filament of gas from the Sun (left), which cools and condenses into planets, with the largest planets in the middle (right). From Thornton Page, *Physics Today* (October 1948).

distance were 100 times greater and the other parameters adjusted accordingly, one star in three might have experienced such an encounter, and "we may, without postulating anything very improbable, suppose our system to have experienced an encounter as close as this. . . ." However, Jeans clearly did not think all these conditions would ensue at one time, and in the end he labeled these occurrences as "somewhat improbable" and systems similar to our own as "somewhat rare," but in general the entire process was not "impossible or very improbable." Given the number of parameters and their uncertainty, Jeans's waffling is not surprising. But he emphasized that no reasonable choice of parameters was likely to alter the result: that only a very few stars have experienced nontransitory encounters. And most important, Jeans stressed, the theory violated no quantitative criterion.

In his classic work *Problems of Cosmogony and Stellar Dynamics* (1919), Jeans discussed the problem in more detail and ended with results even more pessimistic: only one encounter in 30 billion years, a situation so improbable in the present universe as to cast doubt on the validity of the close encounter hypothesis. Pointing out that the parameters were not well known, Jeans concluded that while tidal breakup by a passing star was hardly a likely event, its improbability was not grounds for rejecting the tidal theory. In whatever case one adopted, the solar system seemed to be very exceptional, "and for aught we know may be unique." In his 1923 lecture "The Nebular Hypothesis and Modern Cosmogony," Jeans carried his train of thought one step further, arguing that it was just possible, though not probable, that only the Earth could support life in the universe. "Astronomy does not know whether or not life is important in the scheme of nature, but she begins to whisper that life must necessarily be somewhat rare."

In the hands of Jeans, this whisper soon grew to a crescendo. In both his technical and popular publications by the late 1920s, Jeans (seen at about this time in Fig. 3.3) spread his view far and wide. The numbers varied somewhat, but always present was the basic scenario that the stars are sparsely scattered in space, close encounters exceedingly rare, and the conditions for life very exacting. "All this suggests," Jeans inevitably concluded, "that only an infinitesimally small corner of the universe can be in the least suited to form an abode of life." In his popular works this view of the disruptive approach of stars was vividly drawn, and the rarity of such approaches and their ensuing solar systems was an integral part of this picture – clear even to the public.

For two decades the Jeans tidal theory – with contributions by Sir Harold Jeffreys – was widely accepted, and when the beginning of the end came in 1935 it was once again because of problems with physical principles. This time it was the Americans' turn again, in the form of Henry Norris Russell, who criticized the tidal hypothesis because it could not account for the

Fig. 3.3. Walter Adams, James Jeans (center), and Edwin Hubble, with a model of the 100-inch telescope, April 1931. While the 100-inch helped solve the mysteries of Mars, it could not resolve the problem of planetary systems. Reproduced by permission of The Huntington Library, San Marino, California.

present orbits of the planets. Russell could not see how a close stellar encounter would remove the planets so far from the Sun and give them most of the angular momentum of the system rather than the Sun, which was a thousand times more massive. He also could not see how the planets could condense out of the high-temperature matter ejected from the Sun, an objection given definitive form by Russell's student Lyman Spitzer 4 years later. In their discussion the possibility of other planetary systems played no role, but their fatal objections left science without a workable theory of the origin of the solar system, and by association placed in limbo the idea that such systems were rare.

The 19th-century view of abundant planetary systems was thwarted for decades by the tidal theory of Jeans and Jeffreys. Far from the teleological view of Proctor and others, Jeans's colleague Sir Arthur Eddington asked, "How many acorns are scattered for one that grows to an oak? And need she be more careful of her stars than of her acorns? If indeed she has no grander aim than to provide a home for her greatest experiment, Man, it would be just like her to scatter a million stars whereof one might haply [*sic*] achieve her purpose." To have provided a theoretical underpinning for this startlingly different worldview was no small part of the legacy of James Jeans.

But alas, this worldview had no more claim to objective truth than the 19th-century belief in abundant planetary life, for if the early observational claims for planetary systems at the turn of the century had yielded no definitive result, by 1940 neither had theory solved the problem – nor could it – especially with the departure of spiral nebulae as confirming evidence. The discredited nebular hypothesis had been superseded by the planetesimal hypothesis of Chamberlin and Moulton and then by the tidal theory of Jeans and Jeffreys, only to have Russell and Spitzer overturn the latter, leaving the void. Reviewing the collisional and nebular hypotheses in 1938, Lick Observatory Director Emeritus Robert G. Aitken still saw the development of planetary systems as an "exceptional event." "Exceptional" did not mean unique to Aitken, who pointed out that even if only one star in a million had planets, there would still be 30,000 solar systems in the Milky Way Galaxy – and two million galaxies were within the range of current telescopes.

Aitken, however – an observational astronomer – held out no promise for an observational conclusion. There was in his view "no hope" of a visual sighting. And he had the full weight of his distinguished career in double star astronomy behind him when he pointed out that the gravitational perturbations of Sirius and Procyon were caused by bodies believed to be fully one-third as massive as the primary stars, while our solar system constituted in total only 1/745th of the mass of the Sun. In a pointed analogy, Aitken correctly deduced that our solar system would therefore be much too small to be detected by the gravitational perturbation method from even a nearby star with current techniques. This makes all the more surprising the events that were to follow only 5 years later, events that – justifiably or not – would tranform skepticism into optimism.

3.2 TURNING POINT: 1943–1958

The 15 years between 1943 and 1958 saw a remarkable turning point in the fortunes of planetary systems. It had begun with Russell's criticism of the Jeans–Jeffreys tidal theory, but it was fueled by the revival of a modified nebular hypothesis, by developments in fields as diverse as double star astronomy and the measurement of stellar rotation periods, and – most surprising of all – by insistent claims that planetary systems, or their effects, had been actually observed. Moreover, broader events in the field of cosmology conspired toward change, events that Jeans himself could not ignore.

The implications of the revolution in cosmology of the 1920s and 1930s – a greatly enlarged galaxy, the existence of innumerable "island universes" full of stars, a universe expanding in space and expanded in time – are evident in Jeans's review of the subject of life on other worlds published in 1941. Having given a dim view of the chances of life on Mars and Venus, Jeans turned to the realm of the stars and the origin of planetary systems. He pointed out that under present conditions in the universe the frequency of stellar encounters would be only 1 in 10^{18} years, so that for stars 2 billion years old, 1 star in 500 million might have planets. So far this was his old argument. But it was a sign of the times that he went on to say that though this seemed like a small fraction, in a universe with 10 billion galaxies, each with 100 billion stars, this minute fraction still represented 2 million million stars that might have planetary systems! Statistics – and the new cosmology – had caught up with Jeans, even if only 2000 of these systems might be located in our own galaxy. Straining the definition of "rare," Jeans was forced to conclude that "although planetary systems may be rare in space, their total number is far from insignificant."

By the following year, however, Jeans's view had undergone a much more radical change. In a letter to *Nature* of June 20, 1942, reacting to recent claims of serious dynamic problems arising for the tidal theory assuming that the Sun was approximately its present size at the time of encounter, Jeans asserted that the Sun was most likely comparable in size to the present orbit of Uranus or Neptune when an encounter took place. In a last-ditch effort to save the tidal theory from dynamic objections, Jeans was forced to increase greatly the size of the Sun at the time of supposed planetary formation, a concession that greatly increased its cross section and, by analogy, the cross sections of other suns. Not only did this address the dynamic objections, in Jeans's opinion, it also led to another conclusion: that the total chance of planet formation was now one in six with such a size for the Sun. Thus, "there is no longer any need to strain the probabilities to account for the existence of the planets." And the final conclusion is one hardly expected from Jeans: "A far larger proportion of the stars than we have hitherto imagined must be accompanied by planets; life may be incomparably more abundant in the universe than we have thought." The whole exercise demonstrated the fragility of the argument and the dangers of using equations whose parameters were not well determined. For 25 years Jeans had epitomized the concept of the rarity of life in the universe. Now in the last years of his life he recanted, and his death in 1946 left no substantial heirs to his theory.

Jeans's turnabout was just the beginning, and the cracks opening in the tidal theory in 1941–1942 were to become a breach through which the floodwaters of change would rush in the following year, when strong and independent observational claims were made for the existence of two planetary systems around nearby stars. Many astronomers were quick to draw general conclusions, especially in light of the new observations; as Henry Norris Russell wrote in 1943, "On the basis of this new [observational] evidence, it therefore appears probable that among the stars at large there may be a very large number which are attended by bodies as small as the planets of our own system. This is a radical change – indeed practically a reversal – of the view which was generally held a decade or two ago."

The surprising observational claims for planetary systems in 1943 resulted from the astrometric method, and arose quite naturally out of ongoing astronomic research programs for measuring stellar positions and motions. In particular, photographic measurements of stellar parallax were capable of detecting extremely small shifts in position; the largest parallax was less than a second of arc, and most were much smaller. In the tricky business of angular measurement in astronomy, a process beset by random and systematic errors, this was its most exacting test – and the search for planetary systems would go a step beyond that.

The events of 1943 must be understood in the context of Peter van de Kamp, who would dominate the field of planet detection after the 1930s. A Dutch-born student of Hertzsprung, van de Kamp (Fig. 3.4) was still in high school in 1916, the year that James Jeans first proposed his close encounter theory of the origin of solar systems. But 10 years later he was working in the field of astrometry at the University of Virginia's Leander McCormick Observatory. Van de Kamp had come to the United States in 1923 to help carry out a program of determining the proper motions of stars using photographic techniques with long-focus refractors. It was thus precisely this tradition of stellar positions out of which grew van de Kamp's work, for the same principle is employed in the search for planetary companions. Although in the 1920s and 1930s a planet search was not a specific goal (the goal was merely to measure parallaxes and calculate proper motions), the detection of perturbations in proper motion was a by-product of this parallax method, which was often repeated year after year for the same star. When in 1936 the first perturbation in the proper motion of a star based on photographic observations was announced at McCormick Observatory, it proved to van de Kamp the usefulness of the method of photographic astrometry in the search for low-mass companions. The next year, 1937, he became the director of the Sproul Observatory of Swarthmore College and began a parallax program for the detection of low-mass companions by looking for perturbations in the motions of stars.

While working under van de Kamp at the Sproul Observatory, Kaj Strand was pursuing another line of photographic astrometry: double stars, a field pioneered by Hertzsprung. Visual claims for perturbations in double star orbits had been made in the late 19th century, but in 1943 Strand announced

Fig. 3.4. Peter van de Kamp, pioneer planet hunter of the century, began his search in 1937 at the Sproul Observatory of Swarthmore College and announced a planetary companion to Barnard's star in 1963. He is shown here with the 24-inch Sproul refractor used to make the observations. From van de Kamp, "The Planetary System of Barnard's Star," *Vistas in Astronomy,* 26 (1982), 146. Copyright 1982 with kind permission from Elsevier Science – NL, Sara Burgerhartstraat 25, 1055 KV Amsterdam, The Netherlands.

that he had discovered photographically perturbations indicating a *plane-tary* companion to the star 61 Cygni, famous as one of the first stars to have its parallax measured in 1838. Using photographic observations from several observatories covering the years 1914–1918 and 1935–1942, Strand announced in no uncertain terms, "The only solution which will satisfy the observed motions gives the remarkably small mass of 1/60 that of the sun or 16 times that of Jupiter. With a mass considerably smaller than the smallest known stellar mass (Kruger 60B = 0.14 [solar masses]), the dark companion must have an intrinsic luminosity so extremely low that we may consider it a planet rather than a star. Thus planetary motion has been found outside the solar system." Almost simultaneously, van de Kamp's cousin Dirk Reuyl and Leiden astronomer Erik Holmberg, based primarily on observations made at the Leander McCormick Observatory, announced that they had discovered a planetary companion around the star 70 Ophiuchi. Though they spoke of only a "third body" and not a "planet," the deduced mass for the third body was between .008 and .012 solar masses (compared to .016 for Strand's claimed planet). That this was in the planetary mass range escaped no one.

The reaction to these discoveries was considerable. Immediately upon publication of the results, Russell sat down and wrote an excited account that appeared in the June issue of *Scientific American*. The following month, in an article entitled "Anthropocentrism's Demise," Russell put the results in a broader context for the same magazine. And during the same period, he wrote another article examining from a theoretical viewpoint the physical characteristics of stellar companions of small mass, concluding that "it is well within the bounds of accepted usage to call the new body a planet." A few months later, a discussion erupted in the pages of *Nature* as to the significance of the discovery for theories of the origin of solar systems, in which Jeans himself participated, arguing that with his modification of the previous year, the new discoveries did not affect the status of the tidal theory. But in the final year of his life Jeans declared to a BBC radio audience, "The total number of planets in the whole of space seems to be millions of millions of millions at the lowest." Only 5 years before, one could not have believed it was Jeans speaking.

The observational discoveries were thus widely cited and applauded, even making the pages of *Time* magazine. Fuel was occasionally added to the fire, for example, by van de Kamp's announcement in 1944 of a low-mass companion to Barnard's star and Lalande 21185. Although he concluded that these were probably stellar rather than planetary in mass, they added to the debate over small-mass companions and 20 years later would be the subject of more sensational announcements. Although the claims of Strand, Reuyl, and Holmberg are now generally considered erroneous, they could not be finally disproved for many years; in the meantime, they played a central role in reversing opinions about the frequency of planetary systems.

The increased interest in the subject of low-mass companions generated by these discoveries, as well as its origin in ongoing astrometric programs, is evident in a modest Symposium on "Dwarf Stars and Planet-Like Companions," held in late 1943 under the auspices of the American Astronomical Society. Among the questions discussed was the difference between stars and planets in terms of mass. Russell concluded that an object less than 1/20th the size of our Sun would have a surface temperature of about 700 K and would be invisible under even the best circumstances. The following year, van de Kamp adopted Russell's value of 1/20th (.05) of the Sun's mass "as a conventional borderline between visible stars and the *per se* invisible bodies which we shall designate by the general term *planet*." An agreement on definitions was perhaps a fitting conclusion to the flurry of activity of 1943–1944. But everyone realized that confirmation and further progress in the search for planetary systems would be slow and difficult.

As these observational developments were occurring, news came from Europe that the theoretical vacuum left by Russell's criticism of the tidal hypothesis was beginning to be filled, apparently independently of any knowledge of the observational results announced in the United States. In a Germany engulfed in war, the influential physicist and future philosopher Carl Friedrich von Weizsäcker (1912–) had been working at the University of Strassburg, and in 1944 he published what the first American reviewers called "a significant new paper on the origin of the solar system." A modified version of the nebular hypothesis, it surmounted the traditional problems of that theory by supposing that the greater part of the primordial solar nebula (being composed, in his view, of hydrogen and helium) dissipated into space, carrying with it the angular momentum that had troubled previous nebular theories. The Sun was thus left with little angular momentum, and that of the planets was ascribed to the motion of the original nebula. The planets condensed from the part of the nebula that remained, which included solid particles afloat in a sea of hydrogen and helium. Instead of the accretion process of the Chamberlin–Moulton planetesimal hypothesis, however, von Weizsäcker proposed that the interaction of streams of gas resulted in eddies or vortices that, in turn, yielded planetary orbits roughly analogous to quantized orbits in the Bohr atom.

A measure of the originality of Weizsäcker's theory is that it attracted the attention of scientists of the caliber of Subrahmanyan Chandrasekhar, George Gamow, and many others. It was first made known in the United States in a 1945 review by Gamow and J. Allen Hynek, who concluded that the theory was probably erroneous in details but perhaps correct in some of its qualitative ideas. Among those ideas were the larger mass of the primordial solar nebula; the escape of much of this mass into space before the formation process began, resolving the angular momentum problem; and the physical explanation of the Bode–Titius law of planetary distances. Most important of all, they felt, Weizsäcker "has directed a fresh stream of thought into the long-stagnant pool of theories of planetary origin."

Perhaps sensitized to the issue of the frequency of planetary systems by the emphasis it had been given in Jeans's tidal theory, many astronomers immediately raised this issue, if only in a cursory way. Gamow and Hynek themselves pointed out that "if the planets were formed in this manner, other stars have planetary systems in which similar Bode relations obtain. In any event, if the Weizsäcker theory holds, planetary systems of a wide variety of types must be the rule rather than the exception." With an eye toward confirmation, they also wondered if the process of solar system formation might be going on in the galaxy at present. They concluded that even if it was, a nebular cloud around a star would probably not be visible optically or spectroscopically. Similarly, in a 1948 review of theories of the origin of the Earth, the physicist Thornton L. Page noted of Weizsäcker's theory, "One of the interesting consequences is that the formation of planets should be an extremely common occurrence. Possibly in the process of formation of every star the conditions would be correct to form planets. Thus we might expect billions, if not hundreds of billions of planets in our galaxy, the strong likelihood that life has developed on a million or more of these, the high probability that there are other civilizations of mankind. . . ." This was only the beginning of attempts to make numerical estimates, however uncertain.

One astronomer greatly influenced by Weizsäcker's theory, and destined to make his own contribution to the gathering planetary systems debate, was the Yerkes astronomer Gerard P. Kuiper, whose claim of lichenlike plant life on Mars we examined in the previous chapter. Though he concluded early on that Weizsäcker's theory had to be abandoned in its details, he noted that it had the great merit of "making a fresh start with this difficult problem and of introducing new concepts capable of theoretical analysis." Kuiper first came in the 1930s to the problem of the solar system's origin from the field of observational double star astronomy, and it was in this field that he would make his own contribution to the expanding belief in the abundance of planetary systems. As early as 1935, Kuiper's work on double stars showed that their median separations were about 20 astronomical units, close to the distance of the major planets from the Sun. But it was not until the late 1940s – after the difficulties with the collision hypothesis were well known – that Kuiper connected this result to planetary systems. By 1951 Kuiper was speculating on this basis that "it almost looks as though the solar system is a degenerate double star, in which the second mass did not condense into a single star but was spread out – and formed the planets and comets." On the basis of data showing that 10 percent of binary stars had components with a mass less than 1/10th that of the primary, Kuiper estimated that "planetary systems might be one or two orders of magnitude less frequent than the 10 percent just mentioned," in other words, 1 in 100 or 1 in 1000. "Probably some 10^9 planetary systems occur in the galaxy alone," he concluded. This was orders of magnitude larger than any collision theory predicted, a result that Kuiper noted some found difficult to believe. Referring to the lingering predominance of the theory of Chamberlin, Moulton, Jeans, and Jeffreys, Kuiper recalled, "I announced this result on 4 September 1949 at a regular Sunday broadcast of the University of Chicago Roundtable. I still remember the skepticism of my astronomical colleagues; so strong was astronomical tradition."

Kuiper also addressed the question of whether planetary systems were forming at present, and if so, whether observable celestial objects might be identified with such formation. Spiral nebulae were by this time out of the question as candidates, but he found two other classes of objects promising: a class of young stars known as "T Tauri stars" that often had surrounding nebulae and "Bok globules," dark, circular patches sometimes seen against a bright, nebulous background. Kuiper thus found the empirical study of planetary system formation a possibility in 1956.

Aside from observations of stellar perturbations, the revival of the nebular hypothesis, and double star astronomy, support for planetary systems during this period came unexpectedly from yet another quarter – the study of stellar rotation. It had been known since the time of Galileo that the Sun rotates on its axis, and Frank Schlesinger was the first to actually observe the rotation of stars in 1909. Otto Struve, however, was the pioneer in this field, utilizing stellar rotation as early as 1923 in his thesis to explain the broadening of spectral lines. His earliest published results in 1929 and 1930 established widespread stellar rotation. In his 1930 paper Struve had also noted a sharp slowing down in stellar rotation at the F spectral type, an effect subsequently pinned down more precisely to the F5 spectral type. The cause of this effect was unknown, but in 1930 Struve proposed that it might be related to the breakup of the star into components, forming a binary star. By 1946 he was favorably inclined to the idea that stellar rotation was revealing the possible formation of planetary systems. In his classic *Stellar Evolution* (1950), he proposed that condensation from an interstellar cloud produced a star of the early spectral type with a very high angular momentum and a large mass. As it condensed further it would shed more mass, and (after intermediate stages that included various forms of observed double stars) eventually it would shed its angular momentum by "the formation of a planetary system with a single star having approximately the mass of the sun and devoid of appreciable angular momentum, surrounded by one or more planets at considerable distances, which contain a considerable fraction of the original momentum." Summarizing his ideas of stellar evolution in a figure that included a planetary system as one of its possible outcomes, Struve stressed that the formation of planets was not an essential feature of his hypothesis. By 1952, however, he was not only explicitly attributing the F_5 rotational discontinuity to the formation of planetary systems, he was also suggesting a means for their detection through radial velocity variations. More than 20 years after he clearly accepted axial rotation of stars and the discontinuity in rotation between early and late stars, Struve firmly invoked planetary systems as the explanation. This explanation, in turn, became an important argument in the solar system theory of the British theorist Fred Hoyle and others, and the importance of the stellar rotation argument was spread especially by Struve's student Su-Shu Huang.

Finally, as retired Harvard Observatory Director Harlow Shapley made clear in his popular work *Of Stars and Men: Human Response to an Expanding Universe* (1958), the new cosmology was a continual force in the background favoring abundant planetary systems. That cosmology, he argued, required us to believe that we are not the only life in the universe, and he pointed in particular to three developments. First, the discovery that the nebulae are actually galaxies of stars meant that we have at hand "more than one hundred million million million sources of light and warmth for whatever planets accompany these radiant stars." Second, the expanding universe implied that "a few thousand million years ago . . . the average density in the unexpanded universe must have been so great that collisions of stars and gravitational disruptions of both planets and stars were inevitably frequent . . . at that time countless millions of other planetary systems must have developed, for our sun is of a very common stellar type. . . . Millions of planetary systems must exist, and billions is the better word." If the nebular hypothesis operated – and Shapley stated even at this late date that he believed there is more than one mode of planet formation – then this also implied the existence of a large number of planetary systems. Finally, biochemistry now indicated to Shapley that "whenever the physics, chemistry and climates are right on a planet's surface, life will emerge, persist and evolve." In Shapley's view, the Earth and its life are "on the outer fringe of one galaxy in a universe of millions of galaxies. Man becomes peripheral among the billions of stars in his own Milky Way; and according to the revelations of paleontology and geochemistry he is also exposed as a recent, and perhaps an ephemeral manifestation in the unrolling of cosmic time." This view, poetically expressed by one who had helped to build it, is basically the modern conception of the universe.

In the end, it is important to realize that none of the independent arguments for planetary systems was made with anything approaching deductive rigor. Kuiper's binary star separations, while suggestive of his hypothesis that solar systems were failed double stars, fell far short of proof. While the slowdown in

Table 3.1. *Estimates of frequency of planetary systems, 1920–1961*

Author	Argument	No. of planetary systems in galaxy	No. of habitable planets in galaxy
Jeans (1919, 1923)	Tidal theory	Unique	1
Shapley (1923)	Tidal theory	"Unlikely"	"Uncommon"
Russell (1926)	Tidal theory	"Infrequent"	"Speculation"
Jeans (1941)	No. of stars	10^2	—
Jeans (1942)	> Diameter of Sun	One in six stars	Abundant
Russell (1943)	Observation of companions	Very large	$> 10^3$
Page (1948)	Weizsäcker	$> 10^9$	$> 10^6$
Hoyle (1950)	Supernovae	10^7	10^6
Kuiper (1951)	Binary star statistics	10^9	—
Hoyle (1955)	Stellar rotation	10^{11}	—
Shapley (1958)	Nebular hypothesis	$10^6 – 10^9$	—
Huang (1959)	Stellar rotation	10^9	10^9
Hoyle (1960)	Stellar rotation	10^{11}	10^9
Struve (1961)	Stellar rotation	$> 10^9$	—

Source: Steven J. Dick in J. Heidmann and M. J. Klein, *Bioastronomy: The Search for Extraterrestrial Life* (Berlin, 1991), p. 359, by permission of Springer-Verlag.

stellar rotation examined by Struve might have been the result of the transfer of angular momentum to planetary systems, that transfer might equally have been made to the interstellar medium or other places. And von Weizsäcker's theory merely reverted to the preplanetesimal hypothesis argument that there seemed to be no reason why the process should not be common throughout the universe. For all these reasons, great weight was given to the more direct observation of planetary systems without the intermediary of these theories – and the observations themselves were fraught with difficulty.

Nevertheless, whether accepting the arguments of Kuiper, Struve or von Weizsäcker, independently or in concert, one thing was clear: during this 15-year period planetary systems were returned to their status as a normal outcome of stellar evolution, a comeback evident in Table 3.1. That general idea – opposed by Chamberlin, Moulton, Jeans, and Jeffreys – supported an abundance of planetary systems no matter which of the ascending theories of this period one chose. Moreover, it was once again in consonance with the Copernican principle that our Earth and our solar system were nothing special in the universe.

However doubtful the force of the arguments, there is no doubt of the result of this period that was so tumultuous in science, as well as in world affairs. In 1940 the Astronomer Royal Sir Harold Spencer Jones, echoing Jeans, Eddington, and the conventional wisdom espoused by most astronomers, had written in his volume *Life on Other Worlds* that "life is not widespread in the universe . . . not more than a small proportion of the stars are likely to have any planets at all. With the usual prodigality of Nature, the stars are scattered far and wide, but only the favoured few have planets that are capable of supporting life." In the second edition of 1952, citing von Weizsäcker, he wrote, "On this important problem of the origin of the solar system and of planetary systems in general, there has been a marked change in outlook in the last few years from that of twenty years ago. Astronomers then felt pretty confident that the solar system was something very exceptional; now it appears much more probable that the formation of a planetary system may occur as one of the normal courses of stellar evolution." A turning point had been passed, and there would be no going back for the rest of the century.

3.3 OPTIMISM: OBSERVATION TO THE RESCUE

It is some measure of the difficulty of the search for planetary systems that the Space Age did not bring immediate advances in the problem. Unlike solar system studies, where planetary spacecraft brought immediate and revolutionary progress in our knowledge of the planets, no such prospect was in store for planetary systems. It is true that increased knowledge of our own planetary system provided voluminous data for the refinement of theories on the origin of the solar system, which by the usual gross analogies could be applied to other solar systems. But, although substantial, these refinements changed little the fortunes of planetary systems. Perhaps the largest impact of the Space Age on planetary systems science was the infusion of funds from space agencies such as NASA, which displayed an interest in both observational and theoretical aspects of the subject almost from the beginning, but with delayed results.

We should therefore not be surprised that, while most astronomers in the second half of the 20th century were optimistic about other planetary systems, observational proof of their existence through the 1970s remained entirely dependent on the old astrometric technique. That technique, the results of which remained elusive in many cases, created a public and scientific sensation with the announcement in the 1960s of the detection of several planetary systems. The promise and limitations of this technique, and the difficulties of tackling a problem at the limits of science, may best be seen in the famous case of

Barnard's star. The central figure in the case is Peter van de Kamp, whom we last saw beginning his search for low-mass companions at the Sproul Observatory of Swarthmore College in Pennsylvania in 1937. Such is the long-term nature of the problem of determining perturbations in stellar motions that only now, 25 years later, was van de Kamp beginning to announce results with planetary companions. Barnard's star was a star of 9.5 magnitude, so called after Barnard's discovery in 1916 of its enormous proper motion of about 10.3 arcseconds per year. This meant that it was a close star (the closest known after the Alpha Centauri system), and it was immediately placed on observational programs, including the parallax program at Sproul in 1916–1919. In 1938 van de Kamp had placed it back on the Sproul parallax program with his arrival as director in 1937, and by 1944 he announced a low-mass companion that was stellar in nature. Over the next 20 years, as Kuiper, Struve, and others were predicting an abundance of planetary systems based on their own work, and as theory once again made plausible abundant planetary systems, van de Kamp patiently collected data on Barnard's star and other nearby stars.

There is no doubt that van de Kamp was sensitive to the question of whether low-mass companions were stars or planets, at least since the time of his 1944 article on the subject, an article undoubtedly stimulated by Strand's observational claims of 1943. In a progress report on "Planetary Companions of Stars" in 1956, van de Kamp pointed out that while numerous unseen objects had been detected over the last two decades with masses .05 of the Sun's or greater, it was "extremely likely" that these objects were stars. "There are tentative indications of unseen companion objects with about .01 solar masses *or more,* and these *may* be planetary companions. However, definitive interpretation can hardly be reached at present, partly due to limitations of accuracy," he wrote in that year. In particular, van de Kamp pointed out that the 1943 claims of Reuyl and Holmberg for a planetary companion of 70 Ophiuchi had not been confirmed by Strand as of 1952; and he held out hope only for Strand's claim in 1943 for a companion of 61 Cygni of .016 solar masses, well into the planetary range, as confirmed by two other observers. As for his own program, now almost two decades old, van de Kamp claimed only that the companion to Lalande 21185 was probably a star of low luminosity, and that no satisfactory explanation existed for some perturbations seen in the motion of Barnard's star. Almost two decades after the start of his observational program at Sproul, no one could accuse van de Kamp of rushing to judgment on planetary companions!

This situation was to change in the 1960s, a decade of uproar in this field of astronomy as in so many other areas of American life. First, in 1960 Sarah Lippincott, van de Kamp's colleague at Sproul, announced that the companion of Lalande 21185 had a mass of only .01 that of the Sun. Although this was within planetary range (recall that the 1943 claims for planetary companions to 61 Cygni and 70 Ophiuchi gave them about this same mass), Lippincott's technical article in the *Astronomical Journal* made no mention of the word "planet," and perhaps for this reason her announcement did not raise much of a stir. But van de Kamp's 1963 article, with the mundane title "Astrometric Study of Barnard's Star from Plates Taken with the 24-inch Sproul Refractor," created a sensation. In it he announced the discovery of a companion to Barnard's star with a mass of only .0015 the mass of the Sun, only 1.6 times the size of Jupiter, which he specifically characterized as a planet. Further, he found the distance of the planet from Barnard's star to be similar to that of Jupiter from the Sun, and its surface temperature to be about 60 K compared to 120 K for Jupiter.

Van de Kamp's claim was based on 25 years of photographic observations, using three types of photographic emulsions, and 50 observers, yielding 2413 plates. To the extent that the public was aware of such details, they might have been persuaded by this alone that such an intensive scientific effort must have yielded a definitive result. But they would not have been aware of the subtleties of the technique, which included taking into account a variety of insidious errors that could affect the results. Having taken into account all known sources of error as best he could, van de Kamp found a perturbation in the motion of Barnard's star with a period of about 24 years (Fig. 3.5). In order to come up with an actual mass for the companion, he further had to carry out a "dynamical interpretation" calculation using the mass of Barnard's star. By Kepler's law, once this mass and the period of the orbiting body were known, the mass of the latter could be calculated. It was here that van de Kamp finally came to the figure of .0015 times the mass of the Sun for his new planet: "The orbital analysis leads, therefore, to a perturbing mass of only 1.6 times the mass of Jupiter. We shall interpret this result as a companion of Barnard's star, which therefore appears to be a planet, i.e., an object of such a low mass that it would not create energy by the conventional nuclear conversion of hydrogen into helium."

As with the announcements 20 years before, the reaction to van de Kamp's result was swift. From *Time* to popular science magazines and more sober scientific journals, countless reports of his result hailed the discovery of another planetary system. Independent verification of the result, on the other hand, was more difficult since the observations were very specialized and required decades to reach a result – van de Kamp had been at it for a quarter century. It is not surprising, therefore, that van de Kamp himself was the first to reinforce his own result. In 1969, with 5 more years of photographic measures of Barnard's star, he reiterated his claim that a planetary companion existed around that star, with a slightly revised mass of 1.7 times that of Jupiter. In the same year, van de Kamp proposed an alternative analysis

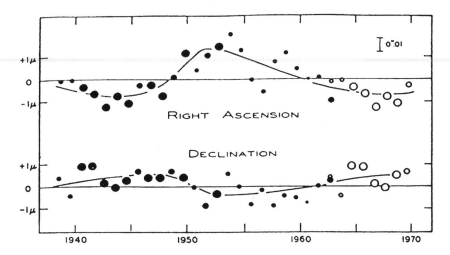

Fig. 3.5. Van de Kamp's data for Barnard's star (1963) represents the classical astrometric method for planet detection showing minute gravitational perturbations of a few hundredths of an arcsecond over a period of decades; plots using the spectroscopic radial velocity method (Fig. 3.9) look similar but need not cover such a long period of time. Used with permission from Elsevier Science Ltd.

of his data that held out the possibility that two planets orbited Barnard's star, with periods of 26 and 12 years and masses of 1.1 and .8 times that of Jupiter.

But trouble was around the corner, and the 1970s saw serious questions raised about van de Kamp's momentous result. In 1973, John Hershey, one of van de Kamp's own students, found that changes made to the Sproul telescope, in particular a change of lens cell in 1949, caused jumps in the data at that point and may have affected the results for Barnard's star. In the same year, George Gatewood of the Allegheny Observatory in Pittsburgh and Heinrich Eichhorn of the University of South Florida concluded "with disappointment," based on an independent analysis of 241 photographic plates, that no perturbations existed in the motion of Barnard's star. Attempting to explain their result, they pointed to the disadvantages of van de Kamp's analytic technique, to changes in the optical system of van de Kamp's telescope over the extended period of time of his study, and to the fact that his claimed perturbation was just "on the verge of significance," a status similar to claimed measurements of parallax before Bessel. A similar analysis by Gatewood published the following year gave the same null result for Lippincott's 1960 claim of a planetary companion around Lalande 21185. Two other studies of van de Kamp's data in 1973 were more favorable to his claim of one or more planetary companions, but a decade after the first announcement by the van de Kamp group, planetary systems were once again in trouble.

Van de Kamp understandably did not take lightly this negation of the main result of his work of 25 years. In order to take the objections into account, especially the finding that changes in the instrument might have affected positional measurements, van de Kamp remeasured his plates on a new machine and included only material from 1950 on. He confirmed the existence of the shorter-period planet, with a mass now .4 that of Jupiter, while the second planet was "less well determined." In 1977, 60 years after the discovery of Barnard's star, van de Kamp took the occasion to reassert his belief in the reality of its planetary companions. In addition to the now familiar scientific defense, the article concluded with a Rembrandt etching on the appearance of Christ to Thomas, with the caption "Blessed are they that have not seen, and yet have believed," suggesting a religious invocation of faith undoubtedly seen by some as not readily transferable to the scientific realm. Van de Kamp's last paper based on new data, published in 1982, again supported the conclusion of two planets around Barnard's star, a conclusion he never relinquished.

The Barnard's star episode was only the most notorious of several claims made for planetary companions by the mid-1970s, all subject to the same limitations of technique and inference. Although the assault on Barnard's star continued to receive the greatest attention, in the field of astrometric perturbations it was not unique and thus could not be written off as a fluke. In an extensive 1975 review of the subject of unseen astrometric companions, van de Kamp listed 17 "well-established perturbations" of stars by unseen companions, including Barnard's star and 3 others with possible planetary companions. Another 14 stars, including the famous 61 Cygni, were listed with "perturbations of provisional, suspected, or uncertain nature."

Not everyone was convinced that even those stars showing well-established perturbations necessarily harbored planets, for this depended on theoretical ideas about the cutoff point for stable hydrogen burning in stars. While in the early 1960s Caltech geochemist Harrison Brown supported the idea of numerous planets by an extension of the "luminosity function" (the distribution of the stars with their visual magnitudes) to low masses, S. S. Kumar argued that all of the objects claimed as planets were probably very-low-mass "degenerate" objects that he termed "black dwarfs." The dividing line between stars and planets remained the subject of vigorous discussion, and the accompanying search for what came to be known as "brown dwarfs" (objects not massive enough to sustain nuclear fusion, generally agreed by now to be below .08 solar mass) remained for decades almost as elusive as the search for planets themselves.

Table 3.2. *Observational milestones in the search for extrasolar planets*

Author	Star	Method
Strand (1943)	61 Cygni	Astrometric (double star)
Reuyl & Holmberg (1943)	70 Ophiuchi	Astrometric (double star)
Lippincott (1960)	Lalande 21185	Astrometric (parallax)
Van de Kamp (1963)	Barnard's star	Astrometric (parallax)
Aumann et al. (1984)	Vega, 40 more	IRAS, IR excess circumstellar disks
Smith and Terrile (1984)	Beta Pictoris	CCD camera image of circumstellar disk
Beckwith (1985)	R Monocerotis HL Tauri	IR speckle interferometry circumstellar disks
Zuckerman and Becklin (1987)	Giclas 29–38	IR excess brown dwarf?
Forrest et al. (1988)	Gliese 569	IR direct image brown dwarf?
Campbell et al. (1988)	9/18 stars	Spectroscopic 1–10 Jup. mass objects (retracted)
Latham et al. (1989)	HD 114762	Spectroscopic brown dwarf?
Lyne and Shemar (1991)	Pulsar PSR 1829 − 10	Radio pulsar timing (retracted)
Wolszczan and Frail (1992)	Pulsar PSR 1257 + 12	Radio pulsar timing
O'Dell et al. (1992)	15 stars in Orion nebula	Hubble Space Telescope circumstellar disks
O'Dell et al. (1994)	56 of 110 stars in Orion nebula	Hubble Space Telescope circumstellar disks
Nakajima, Kulkarni et al. (1995) Durrance and Golimowski (1995)	Gliese 229	IR image of brown dwarf and Hubble Space Telescope image

The only hope for resolution of the question was new (or newly refined) techniques, and by the 1980s those techniques were on the ascendant. As Table 3.2 shows, planet hunting in the 1980s and 1990s belonged not to astrometry, but to methods ranging from ground-based "charged-couple device" (CCD) cameras, infrared speckle interferometry, spectroscopy, and even radio pulsar signals to observations from spacecraft high above the Earth's obscuring atmosphere. The announcement of both ground-based and spacecraft results in the early 1980s fanned hope that the question of the abundance of planetary systems would soon be resolved observationally. A new component of the story was the observation of circumstellar disks as possible evidence of planetary systems in formation rather than the planets themselves. As late as 1995 it seemed that even the new techniques, whether applied to circumstellar disks or to planets, might leave the question of planets around Sun-like stars tantalizingly open.

The new hopes for planet hunting began 25 years after the dawn of the Space Age, with unexpected observations from a spacecraft not specifically designed to search for planetary systems, but rather to survey space for objects at wavelengths just beyond visible light and largely beyond Earth-bound detectors – the infrared region of the spectrum. Beginning in 1983, a team of scientists centered at NASA's Jet Propulsion Laboratory reported that observations made with the Infrared Astronomical Satellite (IRAS) indicated the presence of a cool cloud of solid particles around the bright star Vega in the constellation Lyra. The discovery, one of the first to come from the spacecraft, was made serendipitously while the telescope's detectors were being calibrated. The published paper, entitled "Discovery of a Shell Around Alpha Lyrae," focused on the nature of the shell, which the authors concluded was composed of solid particles at least 1 mm in radius, at a distance of 85 astronomical units (about twice the size of our solar system), heated by the central Sun Vega. The material was believed to be the remnant of the cloud out of which Vega formed. These results, the authors wrote, "provide the first direct evidence outside of the solar system for the growth of large particles from the residual of the prenatal cloud of gas and dust." By early 1984 similar "circumstellar shells" or "protoplanetary disks" had been found around 6 more stars, and by July, with a total of 40 such stars, the shells were being reported as a widespread phenomenon. The discoverers in general were careful to emphasize that planets had not been found; instead, "the presumption is that these rings will eventually condense into solar systems like our own; if so, that makes the Vega phenomenon the first semidirect evidence that planets are indeed common in the universe." In other words, close, but not yet direct evidence of planets.

By late 1984, one of the circumstellar disks found by IRAS had been photographed by a ground-based optical telescope. Employing the new CCD technology in conjunction with the 2.5-m telescope at Las Campanas Observatory in Chile, these observations of the star known as Beta Pictoris produced one of the most famous images in the astronomy of the 1980s (Fig. 3.6), one the public could appreciate more than the infrared excess detected by IRAS. The discoverers, Bradford Smith and Richard Terrile, concluded, "Because

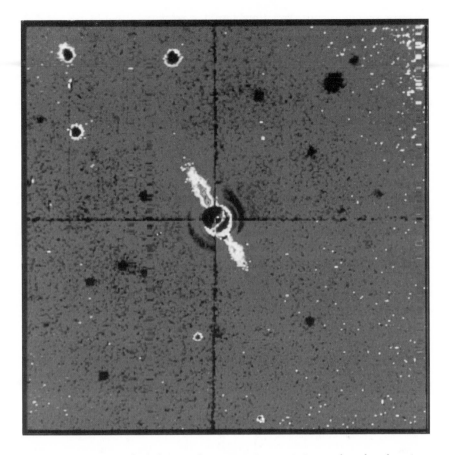

Fig. 3.6. CCD image of a disk around Beta Pictoris (1984). Research undertaken since this discovery indicates that one or more planets may be present in the disk. Courtesy Jet Propulsion Laboratory.

the circumstellar material is in the form of a highly flattened disk rather than a spherical shell, it is presumed to be associated with planet formation. It seems likely that the system is relatively young and that planet formation either is occurring now around Beta Pictoris or has recently been completed." Even as astronomers reported more observations of circumstellar disks, during the 1980s these results remained open to interpretation.

Seemingly closer still was the claim in the late 1980s of indirect detection of actual planets by the spectroscopic method, the fulfillment of an idea foreseen by W. W. Campbell and H. D. Curtis at Lick Observatory in 1905 and the subject of Struve's 1952 proposal. Techniques for determining by spectroscopic means the variations in radial velocity caused by tugs of *stellar* companions had been continually refined over the decades to the level of several hundred meters per second. But only in the 1970s did breakthroughs occur that would reveal such variations at the 10 m/sec level necessary to detect the effect of *planets* tugging on stars, and not until the 1980s would any results be announced relevant to planetary systems. Building on a technique first proposed by Roger Griffin in 1973, with an eye toward the search for extrasolar planets, Bruce Campbell of the University of British Columbia in Canada refined the spectroscopic method using as reference an absorption cell of the highly poisonous gas hydrogen fluoride. In 1988 Campbell and his team announced that after 6 years of observations, 7 of 16 stars examined showed evidence of "long term low level variations" in radial velocity at the level of 25–60 m/sec. These, he concluded, were probably planets with 1 to 9 Jupiter masses, which "probably represent the tip of the planetary mass spectrum."

But by 1990, in a remarkable parallel to the fate of van de Kamp's astrometric claims, cracks had begun to appear in the Canadian assertion that 50 percent of solar-type stars had planetary systems. Campbell himself now showed that at least some of the radial velocity variations were due to astrophysics on the surface of the stars – motions in the stellar material itself – rather than to planetary systems. While planetary systems were still claimed for the other stars, this retrenchment cast doubt on them also. And indeed, by 1995 Campbell's Canadian colleagues Gordon and Andrew Walker and their coworkers announced negative results for their 12-year radial velocity search for planets around 21 solar-type stars.

It might seem that the prospects for planetary systems were at an all-time low, except for those tantalizing circumstellar shells and slowly accumulating progress on the brown dwarf problem. By the late 1980s, several brown dwarf candidates had been announced but not confirmed. Particularly tantalizing was a result announced in 1989 by David Latham and colleagues at the Harvard-Smithsonian Center for Astrophysics and the group led by Michel Mayor and Didier Queloz at the Geneva Observatory. Their observations were the result of a radial velocity survey at precisions of several hundred meters per second, capable of detecting companions as small as 30 Jupiters with periods as long as 10 years. In 1989 they announced a companion to the star known as HD 114762, which Latham and his colleagues reported could have a mass as small as 1/1000th that of the Sun. Not only was this value smaller than the .08 "traditional dividing line between brown dwarfs and stable hydrogen-burning stars," it was also smaller than the "proposed dividing line" between a brown dwarf and a planetary companion. "Thus the unseen companion of HD 114762 is a good candidate to be a brown dwarf or even a giant planet," they concluded, allowing that there was less than a 1 percent chance (depending on the unknown orbital inclination) that this companion was massive enough to burn hydrogen stably. Although the method was incapable of detecting Earth-like planets, the Harvard team argued that any star with a Jupiter-sized planet was likely to harbor a terrestrial planet as well.

Aside from circumstellar disks and brown dwarf candidates, a stunning discovery in a seemingly unrelated branch of astronomy turned up what finally appeared to be the first confirmed planetary system, giving renewed hope to planet hunters everywhere. In a bizarre twist to the planet detection story, this planetary system was found in 1992 circling a pulsar, one of a class of extremely dense, rapidly rotating neutron stars believed to be born from supernova explosions – long past the evolutionary stage that might harbor conditions for life. The technique still depended on the gravitational tug of a planet on the star, but with a considerable difference from the classical method: because pulsars emitted extremely accurate clock-like pulses at radio wavelengths, the tug could now be measured by timing pulse differences amounting to only a few thousandths of a second as the pulsar was tugged first one way, then another by the circling planet – making this the most sensitive (if unexpected) planet detection technique known. Of the three planets discovered around the pulsar, two had three times the mass of the Earth, and one had the mass of the Earth's Moon. Almost all astronomers agreed that the evidence for pulsar planets was "completely irrefutable." Had Struve still been alive, he would have marveled that planets had seemingly been found, not around those slowly rotating stars that he believed in the 1950s had given up their angular momentum to the planets, but around the most rapidly rotating objects in the universe.

By the early 1990s, then, pulsar planets, numerous circumstellar disks, and several brown dwarf candidates kept alive the hopes of exasperated planet hunters everywhere. The Hubble Space Telescope's detection in late 1993 of "protoplanetary disks" around 56 of 110 young stars in the Orion Nebula (Fig. 3.7) added further fuel to the fire. But were extrasolar planets common or not? The question was still open, and it was not yet clear whether the evidence would eventually go the way of 61 Cygni, 70 Ophiuchi, and Barnard's star, or whether it was revealing the tip of a planetary iceberg. While some pointed out that pulsar planets were hardly the kind of planets of interest for life, others argued that if planets could form around pulsars, they could form anywhere. It remained uncertain whether circumstellar shells had formed or would ever form planets. And even if the existence of the brown dwarf candidates was confirmed, they were not the same as planets. While the direct imaging of a brown dwarf in 1994 (Fig. 3.8) finally provided graphic evidence of the existence of low-mass objects, the early 1990s still found the search for planetary systems in a curious limbo. It was a maddening parallel to early-19th-century attempts to determine stellar parallax, a measurement that John

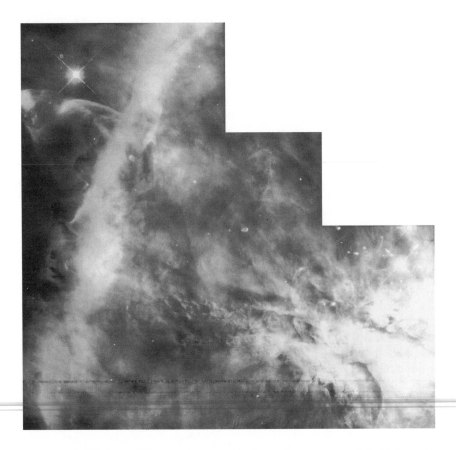

Fig. 3.7. Hubble Space Telescope image, taken December 29, 1993, of the Orion nebula, a star-forming region 1500 light years distant. Some of the small knots of matter
are believed to be protoplanetary disks, or "proplyds," that might evolve into planets.
By permission of NASA and C. R. O'Dell/Rice University.

Herschel characterized on behalf of the hapless parallax seeker as "continually hovering just beyond the limits of his distinct apprehension, and so leading him on in hopeless, endless, and exhausting pursuit."

This situation changed with the remarkable events of 1995 and 1996, years
that will surely be seen as a turning point in the search for planetary systems. In October 1995 Michel Mayor and Didier Queloz of the Geneva Observatory in Switzerland announced their discovery of a Jupiter-sized planet
in orbit about the Sun-like star 51 Pegasi. Finally, the radial velocity method
seemed to have paid off (Fig 3.9). Those who knew their history could be

Fig. 3.8. First unambiguous detection of a brown dwarf star, observed (left) in 1994 with adaptive optics using the 60-inch reflector on Palomar Mountain and confirmed (right) in 1995 with the Hubble Space Telescope. The brown dwarf has a mass only 20–50 times that of Jupiter and circles the star Gliese 229, which is 18 light years from Earth. The brown dwarf circles its star at about the distance at which Pluto orbits the Sun in our solar system. Left, courtesy T. Nakajima (Caltech), S. Durrance (Johns Hopkins). Right, courtesy S. Kulkarni (Caltech), D. Golimowski (Johns Hopkins), and NASA.

forgiven for being skeptical of yet another planet claim. Indeed, there were skeptics. But not only was this finding confirmed within a few weeks by Geoff Marcy and Paul Butler of San Francisco State University, in January 1996 Marcy and Butler announced two more planets, one around 47 Ursae Majoris and another around 70 Virginis, both 40 to 80 light years away. Before the year was out, a total of eight planetary systems had been announced around seven Sun-like stars using this method (Table 3.3). All of this was the result of many years' previous work, patiently gathering data using the spectroscopic radial velocity technique. Mayor and Queloz had been observing 142 Sun-like stars since 1994, working at a precision of 13 m/sec. Marcy and Butler had been observing 120 solar-type stars with Lick Observatory's 120-inch telescope since 1987, in the end achieving a precision of 3 m/sec. The planets around 51 Pegasi, with its 59 m/sec tug, and around 70 Virginis, with its enormous 318 m/sec tug, were easy pickings even though for technical reasons the interpretation of the 51 Pegasi data remained contentious for several years. But many more planets were expected to be found as other groups made use of the new technique.

At the same time, astrometry was not completely out of the planet game either. In June of 1996, George Gatewood, the same Allegheny Observatory

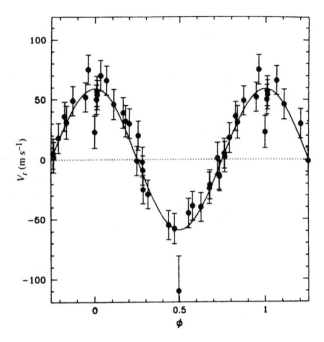

Fig. 3.9. Evidence of a planet with about one-half Jupiter's mass around 51 Pegasi. The sinusoid represents a line-of-sight variation in the motion of the star of ±59 m/sec as the star is tugged one way and then another over 4.2 days by the inferred planet. The spectroscopic "Doppler" method works best for planets close to a parent star, while the astrometric method depicted in Figure 3.5 is optimal for planets more distant from their star. Reprinted with permission from Michel Mayor and Didier Queloz, "A Jupiter-Mass Companion to a Solar-Type Star," *Nature*, 378 (1995), 355–359, copyright 1995 by Macmillan Magazines Limited.

astronomer who in 1973 had failed to confirm van de Kamp's claims for Barnard's star, announced finding planetary companions around Lalande 21185, the very star to which van de Kamp's colleague, Sara Lippincott, had given planetary aspirations in 1960. Gatewood's claim stood out from other claims for many reasons: Lalande 21185 was the sixth closest star to our Sun at only 8 light years away; the observations indicated a Jupiter-sized planet in a nearly circular orbit at about the distance of Saturn in our own solar system; and further perturbations indicated that the system had possibly as many as three planets, and they were detected astrometrically, not spectroscopically. In other words, this was the first solar system similar to our own rather than harboring massive planets very close to a parent star. As such, these planets were primary candidates for the first direct imaging of a planet beyond our

Table 3.3. *Extrasolar planets in order of discovery*

Star	Date announced	Discoverer	Distance from Earth (light years)	Minimum mass (Jupiters)[a]	Distance from star (Earth–Sun = 1)	Period (days)	e[b]
HD 114762[c]	1989	D. Latham	90	10	.3	84.04	.38
PSR 1257+12	Jan 1992	A. Wolszczan	1600	.015E	.19	25.34	0.0
				3.4E	.36	66.6	.0182
				2.8E	.47	98.2	.0264
PSR B1620−26	1993	D. Backer/ S. Thorsett/ S. Sigurdsson/ et al.		2.8E	> 10?		
51 Pegasi	Oct 1995	M. Mayor/ D. Queloz	50	.47	.05	4.2	0.0
70 Virginis[c]	Jan 1996	G. Marcy/ P. Butler	78	6.6	.43	116	0.40
47 Ursae Majoris	Jan 1996	P. Butler/ G. Marcy	46	2.8	2.11	1090	.03
55 rho 1 Cancri	Apr 1996	G. Marcy/ P. Butler	45	.84	.11	14.6	.05
				5	7?	5500?	.05
Lalande 21185[d]	Jun 1996	G. Gatewood	8.2	.9		5.8 years	
				1.6		30 years	
Tau Boötis	Jun 1996	P. Butler/ G. Marcy	50	3.87	.0462	3.3	0.0
Upsilon Andromedae	Jun 1996	P. Butler/ G. Marcy	54	.68	.057	4.6	0.0
16 Cygni B	Oct 1996	B. Cochran/ A. Hatzes/ G. Marcy/ P. Butler	85	1.5	2.2	810	.63
Rho Coronae Borealis	Apr 1997	R. Noyes et al.	50	1.13	.23	39.6	.028

[a] Pulsar planets are expressed in Earth masses.
[b] e = eccentricity, where 0 is perfectly circular.
[c] Possible brown dwarf.
[d] Unconfirmed.

solar system. Even though the planet would be a billionth the brightness of its parent star, this was a feat that might be achieved at near-infrared wavelengths by the Hubble Space Telescope or from the ground using adaptive optics. Until that happened, the system was considered unconfirmed.

In 1997 a team of scientists at three institutions in the United States announced a possible planet around the star CM Draconis using the photometric technique of measuring light differences as the planet eclipsed its star. Confirmation of this report would prove the success of the last of the three classical methods of planet detection.

A glance at Table 3.3 shows a curious phenomenon, aside from the rapid pace of discovery. Some of the planets formed much closer to their host star than expected, in fact so close that theorists had to scramble to explain how they could have formed there at all. From the limited sample of discoveries, three groups of planets could already be distinguished. The 51 Pegasi type are hot Jupiters – very close to their star, with periods of a few days; the planets of 55 Cancri, Tau Boötis, Upsilon Andromedae, and Rho Coronae Borealis fall into this category. The planets in very eccentric orbits are HD 114762, 70 Virginis, and 16 Cygni B. Jupiter-sized worlds in more distant orbits that are close to circular include 47 Ursa Majoris and Gatewood's planets around Lalande 21185. Considerable controversy remained, however, regarding the objects in very eccentric orbits. Some astronomers suggested that highly eccentric orbits were an indication that these objects were brown dwarfs rather than planets, since planets forming by accretion yield nearly circular orbits, at least in our solar system, while brown dwarfs forming by a collapsing gas cloud may yield eccentric orbits. Much was at stake in this argument, including whether or not the first extrasolar planet around a Sun-like star had already been discovered in 1989, in the form of HD 114762. With more announcements undoubtedly to come, there was the real prospect that by early in the 21st century a classification of planetary systems might be within grasp.

Thus, within the short space of a year or two at the end of the century, when some were beginning to think the question might never be resolved, the scales were tipped sharply in favor of abundant planetary systems. Though some cautioned that a single planet around a Sun-like star did not constitute a planetary system, and though everyone cautioned that none of these planets appeared habitable based on their orbital characteristics, one of the holy grails of the extraterrestrial life debate had finally been achieved. Moreover, and perhaps just as important, knowledge of the nature of circumstellar disks increased rapidly. Hubble Space Telescope images showed that the disk around Beta Pictoris is relatively thin (less than 1 billion miles), increasing the possibility that accretion of large bodies has already occurred. A detected warping of the disk and a depletion of dust in part of it also suggested that one or more planets may have formed in it. Images of dust and gas surrounding much younger stars (including one known as MWC 480, only 6 million years old compared to 100 million years for Beta Pic) gave evidence of planet formation in earlier stages, and the analysis of circumstellar disks in general held out every hope of revealing a menagerie of planetary systems in various stages of formation. The implication was clear: "Because the conditions necessary to make planets appear to be common," Steven Beckwith and Anneila Sargent wrote in 1996, "we believe that planetary systems themselves are also common."

All of these scientific developments had a political result too. As a result of the flurry of activities in planetary systems science over the decade leading to 1995, planet hunting was capturing increased attention from NASA, which contemplated a systematic and comprehensive program of its own. Although the American space agency had supported a low level of study from the 1970s, the call to action came in 1985 when NASA's Solar System Exploration Division established a Planetary Astronomy Committee whose scope included not only the study of our solar system but also the search for planetary systems. In the same year, the Space Science Board of the National Academy of Science directed its Committee on Planetary and Lunar Exploration (COMPLEX) to widen its scope to planetary systems. By 1988 they had decided that preparations for the search should begin in earnest; accordingly, the Solar System Exploration Division created a Science Working Group "to formulate a strategy for the discovery and study of other planetary systems." During a workshop in Houston in early 1990, a three-phase program known as "Toward Other Planetary Systems" (TOPS) was formulated, to begin with ground-based observations, followed by a space-based effort and study of the discovered planetary systems. By 1995 TOPS had been transformed into ExNPS (Exploration of Neighboring Planetary Systems), and the effort to find planets was integrated into NASA's Origins program. The reports generated by these studies discussed the full range of planet detection problems in a practical and programmatic way, defining the common understanding of such terms as "planet," "brown dwarf," and "star," and quantifying them in graphic form (Fig. 3.10).

While the participants contemplated mainly an indirect search in the early stages of the project, with the advance of space technology, even the direct imaging of extrasolar planets was not out of the question. In fact, the detection of Earth-like planets, and even the study of biogenic components of planetary atmospheres, became a long-term goal of NASA's Origins program under its administrator, Daniel Goldin. Ground-based and space-based platforms using the techniques of astrometry and radial velocity, interferometry, and gravitational lensing were all part of the planned program. The space agencies of other countries contemplated their own programs. With these activities, it was clear that the search for planets beyond the solar system was no longer a solitary occupation, as it had been with van de Kamp, but was developing into the discipline of planetary systems science proclaimed by its participants.

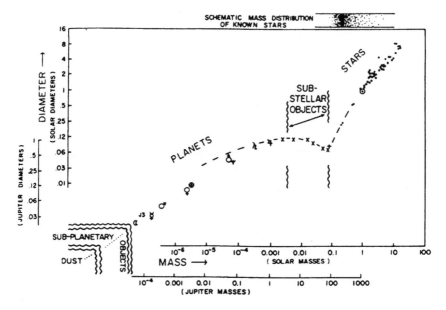

Fig. 3.10. Diagram of diameter versus mass showing classes of objects in the search for planetary systems, ranging from dust (lower left) to planets, brown dwarfs, and stars (upper right). Reprinted with permission from *Strategy for the Detection and Study of other Planetary Systems and Extrasolar Planetary Materials.* Copyright 1990 by the National Academy of Sciences. Courtesy of the National Academy Press, Washington, D.C.

As had happened in planetary science, until 1995 indefinite results in planetary systems science invited a broader role for preconceived notions. Planet searchers, driven by considerations of stellar evolution, a revived nebular hypothesis, the Copernican principle of mediocrity, and simple analogy certainly expected other planets to exist and fervently hoped to find them. But the search for planetary systems should not be seen in the same class as the search for Lowell's canals. Unlike the canals, stellar perturbations by planetary bodies were not a result of direct visual perception, but rather of a whole series of steps, from the detector, to the measurement process, to statistical analysis of the data. The high stakes and fervent hopes notwithstanding, in the end the search for planets outside the solar system is another example of science attempting to function at its outermost limits. Instead of parallax, astronomers looked for *variations* in proper motions through the parallax technique; instead of a few *tenths* of a second of arc, they attempted to extract from their instruments a few *hundredths* of a second of arc. Instead of stellar companions causing velocity variations in the primary star at the level of *kilometers per second,* they now searched for planets at *meter per second* variations. And where once tenths of a magnitude pushed the limits of photometry, now differences of hundredths of a magnitude were required. Yet in the end – and this is a significant lesson – the search achieved a result that approached consensus.

Throughout all of the debate, there should be no doubt of the ultimate aim of the field now known as planetary systems science. Although the search for extrasolar planets arose naturally as part of ongoing research programs in astronomy, in the 20th century this search became an important component of what may be viewed as a new cosmological worldview. This cosmology assumed that planetary systems were common, that life has developed on many of those planets, and that intelligence may have evolved to the point where we can communicate by radio waves. The goal was to test the validity of this worldview, a point repeatedly stressed by participants. Seen in this light, the search for planetary systems has taken on a vigorous life of its own, the first and most crucial of many tests of this new cosmology, conferring on it a role of even more significance to the history of science.

4

EXTRATERRESTRIALS IN LITERATURE AND THE ARTS
The Role of Imagination

With infinite complacency men went to and fro over this little globe about their affairs, dreaming themselves the highest creatures in the whole vast universe, and serene in their assurance of their empire over matter. . . . Yet across the gulf of space minds that are to our minds as ours are to those of the beasts that perish, intellects vast and cool and unsympathetic, regarded this earth with envious eyes, and slowly and surely drew up their plans against us.

H. G. Wells (1897)

Space . . . the final frontier. These are the voyages of the starship Enterprise. Its five-year mission: to explore strange new worlds, to seek out new life and new civilizations – to boldly go where no man has gone before. . . .

Star Trek (1966)

It is, I admit, mere imagination: but how often is imagination the mother of truth?

Sherlock Holmes

As flies to wanton boys are we to th' gods.

Shakespeare (1605)

One of the most surprising developments in the saga of extraterrestrial life in the 20th century was the spread of the idea from science to the realm of the arts. Despite the popular appeal of the idea in the past, no one could have foreseen that this subject, of all the rich variety of scientific subjects, would be transformed into one of the universal themes of literature. That this is precisely what happened is some measure of how deeply felt and firmly ingrained was the alien concept in the human mind. And with this important step, it became even more entrenched, for now the concept evoked not only an intellectual but also an emotional response.

During the 20th century the concept of the alien became a leitmotif of that young genre known as "science fiction," developed successively in literature and film and culminating in some of the most popular movies of all time, including *2001: A Space Odyssey* (1968), *Close Encounters of the Third Kind* (1977), *E.T.: The Extraterrestrial* (1982), and *Independence Day* (1996). For vast segments of the public, the idea of extraterrestrial life entered their lives not as science or philosophy, but as a fictional account that gave readers, and eventually viewers of film, an emotional experience with the alien unfettered by bothersome scientific technicalities. Science stimulated the literary

106

imagination, and literature and film, in turn, stimulated the popular imagination, to such an extent that a history of the alien in science fiction is necessary to a full understanding of why extraterrestrials came to have such a hold on the popular mind. Moreover, a surprising number of scientists were influenced by what we might call "alien literature" and even produced some of the best of it themselves, forming an increasingly symbiotic relation with science in the second half of the 20th century.

In this chapter we document the origin, development, and coming of age of the alien theme in science fiction, with an eye toward its role in establishing the alien in popular culture as well as in the scientific imagination. So pervasive did the theme become that our approach cannot be encyclopedic. Instead it will be sufficient to show from a few examples the broad patterns in the development of the alien theme and the interaction of the alien in literature with science and the extraterrestrial life debate.

4.1 THE INVENTION OF THE ALIEN: VERNE, WELLS, AND LASSWITZ

It is a remarkable fact of history that only in the last third of the 19th century did extraterrestrials enter the realm of literature. Though Lucian's armies of the Sun and the Moon clashed in the second century, though Kepler's remarkable imagination placed thick-skinned Selenites on the Moon, and though Voltaire had mile-high Sirians in his *Micromégas,* the true extraterrestrial alien, replete with its own physical and mental characteristics, is a relatively recent invention. Why should this be when the concept of extraterrestrials is so old? It is the thesis of this section that the birth of the alien in literature is closely tied to late 19th-century science, especially evolutionary theory, astronomy, and the plurality-of-worlds tradition. We shall see these influences repeated in three traditions: the French, German, and British, while the imaginative science of the American astronomer Percival Lowell fueled these traditions after their birth.

The history of the alien in literature may be approached from many sides: the history of science fiction, the history of the extraterrestrial life debate, the history of science, and even the history of religion. To contrast only two of these, it is immediately evident that if the alien came late in the history of science, it came remarkably early in the history of science fiction. Though precursors to science fiction literature are well known, notably Mary Shelley's *Frankenstein* (1818), three figures are most often associated with its birth: Jules Verne in France, H. G. Wells in England, and, for those who probe a bit deeper, Kurd Lasswitz in Germany. If one would expect in Proctor's England that an H. G. Wells would arise to send Martians to invade the Earth, one would expect no less in Flammarion's France and even in a Germany still reeling from Kant's philosophy. It is curious, then, that while Wells and Lasswitz follow this expectation, Jules Verne does not, a circumstance that gives us a clue to the invention of the alien in literature.

We may begin in France, for it is here that the remarkable imaginations of Camille Flammarion and Jules Verne flourished and here that the no less remarkable but now more obscure works of J. H. Rosny the Elder were penned. As we have seen in Chapter 1, in 1862, at the age of only 20, Flammarion's *La pluralité des mondes habités* (Plurality of Inhabited Worlds) appeared, a factual account of the possibility of life on other worlds that was to go through 33 editions by 1880, as well as numerous translations and reprintings well into the 20th century. This work was fully in the plurality-of-worlds tradition, as was his *Les mondes imaginaire et les mondes reels* (Real and Imaginary Worlds), published 2 years later. But Flammarion soon spilled beyond the bounds of science when in 1872 he wrote his *Recits de l'infini*, translated in 1874 as *Stories of Infinity,* which included three tales treating other worlds: "Lumen," "The History of a Comet," and "In Infinity," in which a disembodied spirit travels through the universe, reincarnating on other worlds in forms that might have resulted from evolution under alien conditions. Flammarion's "boldest scientific romance," *La fin du monde* (translated as Omega: The Last Days of the World) (French, 1893–1894; English, 1897), has been seen as a precursor to the British philosopher Olaf Stapledon's *Last and First Men* (1930), a seminal work of science fiction that we shall discuss in the next section. In both fact and fiction, then, the alien pervaded Flammarion's work, and through him spread into the intellectual life and popular culture of France.

Jules Verne (1828–1905) was not as single-minded as Flammarion, nor did he have a scientific background. He was already 34 when he launched his career with *Cinq semaines en balloon* (Five Weeks in a Balloon) in 1863, 1 year after Flammarion's *Pluralité.* There followed in rapid succession the *Voyage au centre de la terre* (Journey to the Center of the Earth, 1864), *De la terre a la lune* (From the Earth to the Moon, 1865), *Autor de la lune* (Around the Moon, 1870), and *Vingt mille lieues sous le mers* (Twenty Thousand Leagues under the Sea, 1870), among others. Like Flammarion, Verne lived in Paris and was familiar with the young astronomer's work. Yet, throughout a writing career that lasted until his death in 1905, Verne never wrote a novel that made significant use of aliens.

Verne's brief discussion of extraterrestrials in his novel *Around the Moon,* however, provides a clue as to why he never wrote an alien novel. The question of life on the Moon, one of the characters says, "requires a double solution. Is the Moon habitable? Has the Moon ever been inhabited?" Verne decides the Moon is not habitable because of "her surrounding atmosphere certainly very much reduced, her seas for the most part dried up, her insufficient supply of water, restricted vegetation, sudden alternations of cold and heat, her days and nights of 354 hours." Moreover, Verne's character adds decisively as they approach the Moon, viewed from 500 yards there is no sign of motion and thus no sign of life. Verne concluded, however, that the Moon was inhabited in the past, "for nature does not expend herself in vain; and a world so wonderfully formed for habitation must necessarily be inhabited." And inhabited "by a human race organized like our own; that she has produced animals anatomically formed like the terrestrial animals; but I add that these races, human or animal, have had their day, and are now forever extinct!" The Moon, Verne had his character say, had grown older more quickly than the Earth, its rarefied atmosphere now rendering it uninhabitable, as someday the Earth would be.

Here we have our clue to Verne's relative neglect of extraterrestrials in his stories. Though it has been said that Verne wanted to do for science and geography what Dumas had done for history – to excite through the power of literature an emotional response to otherwise mundane facts of everyday existence – by comparison to Flammarion and others, his imagination was constrained by science. Although willing to grant that the evolution of worlds would yield an inhabited Moon in the past, Verne required evidence for a Moon presently inhabited. With his fertile imagination the French author would venture around the world and into space, but he never went beyond the Moon to Mars or took up Flammarion's passion for the alien.

Verne's imagination paled by contrast to that of his younger countryman J. H. Rosny the Elder (1856–1940). Though now obscure, Rosny was the one rival Verne had to contend with in France, and it is generally considered that Verne's decline in the 1880s and 1890s coincides with the rise of Rosny's remarkable works. Undoubtedly influenced by Flammarion, his very first work, "Les xipehuz" ("The Shapes," 1887) dealt fictionally with wholly alien themes. Not for Rosny a discussion of life on the Moon; rather, he tackled the nature of life itself, though with little literary merit at this early stage in his career. In "The Shapes" mysterious translucent creatures appear on Earth, threatening the existence of humanity. Unlike anything in terrestrial experience or likely to emerge from terrestrial evolution, the creatures are living minerals unable to communicate with humans. Although they are eventually destroyed, Rosny succeeds in vividly conveying the idea that life in the universe may be completely unlike life on Earth. This theme of strange life, though detached from its extraterrestrial associations, is continued in "Un autre monde" ("Another World," 1895) and La mort de la terre (The Death of the Earth, 1910), the latter dealing with ferromagnetic life.

In his conception of life and intelligence totally incompatible with terrestrial life, Rosny was almost a half-century ahead of his time, and one wonders how the alien theme might have developed had not the world become almost

Fig. 4.1. A pioneer in alien literature: Kurd Lasswitz, author of *Auf Zwei Planeten*.

obsessed with the myth of a Martian civilization. As we saw in Chapter 3, this idea had begun with Schiaparelli's observations in 1877 but captured the popular and literary imagination only with the publication of Lowell's *Mars* in 1895. It was in November of that very year that the German philosopher and historian Kurd Lasswitz (1848–1910) began his novel *Auf Zwei Planeten* (On Two Planets, 1897), in which the Martians traveled to Earth and accidentally came face to face with its inhabitants. A teacher of mathematics, physics, and philosophy at Gotha, a town southwest of Berlin, Lasswitz (Fig. 4.1) was steeped in the traditions of science and German idealism. Best known for his biography of the physicist and philosopher Gustav Fechner (which appeared as he was writing his Martian novel) and his *Geschichte der Atomistic von Mittelalter bis Newton* (History of Atomism from the Middle Ages to Newton, 1889–1890), he is often given the title "father of German science fiction," holding the place in Germany that H. G. Wells does in England and Verne and Rosny in France. Beginning as early as 1878, Lasswitz began to examine the basis of fiction about science, arguing that it satisfied a basic human need and was a legitimate form of art.

What led Lasswitz to write his novel is nowhere explicitly stated, but we can make a plausible case that it was a combination of his interest in extraterrestrials and his desire to see the improvement of terrestrial society. It is unlikely that Lowell provided the initial stimulus for Lasswitz's novel, but the latter may have known of Schiaparelli's work and perhaps of Flammarion's *La planète Mars* (1892), and internal evidence indicates that he must have learned of Lowell's *Mars* as he wrote the novel. Whatever the proximate causes, it is clear that Lasswitz was affected by the earlier plurality-of-worlds tradition. In an article on extraterrestrial life written near the end of his life, he said that ever since science had made the Earth a planet and the stars suns, "we have not been able to lift our gaze to the starry firmament without thinking, along with Giordano Bruno, that even on those inaccessible worlds there may exist living, feeling, thinking creatures. It must seem absolutely nonsensical indeed that in the infinity of the cosmos our Earth should have remained the only supporter of intelligent beings. The rational order of the universe demands that there should necessarily even be infinite gradations of intelligent beings inhabiting such worlds."

But what set Lasswitz to writing his novel was more than a belief in extraterrestrials; it was a belief that they could be used to illuminate some of his strongly felt ideas about society and the important role of science and technology in it. A better society, he felt, was not only within reach but had already been achieved in some part of the universe. In this endeavor, the general philosophical influences on Lasswitz were Kant and, more directly, Fechner, from both of whom he took the idea of free moral will that pervades the novel.

On Two Planets portrays Martians who have arrived in advance of a larger Martian expedition, established a solar-powered space station "hovering" above the Earth's North Pole, and constructed a base at the Pole itself, where they are discovered by a German balloon expedition. The canaled Mars they left behind was straight out of Lowell. It was not desperation that brought the Martians to Earth but curiosity: with their more transparent atmosphere their telescopes had been able to detect cities on Earth, and after repeated attempts at communication by signal, they had discovered how to control gravitation and travel to Earth.

Lasswitz's Martians, having mastered technology in order to survive the scarcity of water on their planet, were advanced in every way over their terrestrial counterparts: "Mathematics and natural science had reached a climax in their development which looms before us humans as a distant ideal. . . . The conditions of Mars favored the development of culture and civilization to a much higher degree than the conditions of earth." Although the Martians were humanoid and capable of interbreeding, they "had large heads, very light, nearly white hair; shining, powerful, piercing eyes." More important, they had reached a higher ethical level than their terrestrial counterparts.

Free from instincts and desires, they had found contentment in following their "immortal philosopher" Imm.

In biological, social, and technological aspects, Lasswitz thus adopted an evolutionary universe, showing the influence of Kant, Fechner, and Lowell. Employing both the Earth and Mars as settings, he uses the situation to explore the relation between two cultures and the problems of the improvement of society. When the superior Martians place the Earth under a benign protectorate, the Earthlings find their human dignity at risk and eventually rebel. When they are finally able to seize the polar base and space station, a treaty is negotiated and communication between the two civilizations is limited to light signals. An equality between the two civilizations is achieved, and as the story ends there is hope that the new planetary order may lead to the long-sought Utopia.

Although the lack of an English translation before 1971 prevented much impact in England or America, the influence of *Auf zwei Planeten* in Germany is witnessed by its sales of almost 100,000 copies and Wernher von Braun's statement at the beginning of the English edition that "I shall never forget how I devoured this novel with curiosity and excitement as a young man." Within 10 years of its publication in 1897, *Auf zwei Planeten* had been translated into Swedish, Norwegian, Danish, Dutch, Spanish, Italian, Czech, Polish, and Hungarian.

As Karl Guthke has pointed out in *The Last Frontier,* if Lasswitz's story left a legacy of hope, H. G. Wells's Martian fantasy left a legacy of fear. *The War of the Worlds,* serialized in 1897 just as Lasswitz's novel was emerging in Germany, and first published in book form in 1898, is striking in its originality: Wells not only led the first invasion from outer space but did so in fine literary style. The influence and appeal of the work are apparent not only in its many editions and in the well-known effect of the radio play adaptation in 1938, but also in innumerable imitations that have followed in the century since. As Arthur C. Clarke has written, the Menace from Space was virtually unknown before Wells but has become all too common since.

How do we explain the origin of this prototype of alien literature at this time and place? Wells (Fig. 4.2) was a man of his time in many ways, not least in his championship of Darwin's theory of evolution. As Fechner was Lasswitz's great influence, the biologist T. H. Huxley was Wells's. Huxley, the 19th century's greatest champion of Darwinian evolution, taught Wells from about 1883 to 1886 at the Normal School of Science (later known as the Royal College of Science) in London. Huxley's evolutionary viewpoint is pervasive in Wells's early writings in science and science fiction, and it forms the broad background to the advanced beings in *The War of the Worlds.*

Moreover, it is clear that Wells was familiar with the plurality-of-worlds tradition. He knew that the debate was at least as old as the 17th century, as

Fig. 4.2. H. G. Wells. Courtesy of the Rare Book and Special Collections Library, University of Illinois at Urbana–Champaign.

his prefatory quotation from Kepler indicates: "But who shall dwell in these worlds if they be inhabited? . . . Are we or they the Lords of the World? . . . And how are all things made for man?" As a widely read man of his time, he could not fail to have known of Richard Proctor's *Other Worlds Than Ours,* which by then had gone through at least seven English editions. He may have known of Huxley's own belief on the subject, as well as Flammarion's *Pluralité,* though no English edition had appeared. Perhaps most important, it was no accident that Wells's invaders came from the Mars of Percival Lowell, whose Martian canal controversy had by now reached England.

We know in particular that Wells was interested in life beyond the Earth as early as 1888, when he spoke before the Debate Society at the Royal College of Science on the question "Are the Planets Habitable?" In June 1894 he wrote "The Living Things That May Be," a review of J. E. Gore's *The Worlds of Space,* in which he faulted Gore for lack of imagination and suggested the possibility of silicon life. In December of the same year he followed up on this theme in "Another Basis for Life." "Very attractive is the question of whether life extends beyond the limits of this little planet of ours," Wells wrote, citing the recent work of the astronomer Sir Robert Ball on conditions for life in space. Finally, in April 1896, 1 year before *War of the Worlds* began to appear serially, his essay "Intelligence on Mars" appeared, in which he argued that if Martians exist, they are very unlike us.

In 1893 Wells began to turn out short stories in earnest, and as early as 1894 these stories showed a penchant for strange life forms, although all were on Earth. By 1897 Wells's fascination with the extraterrestrial is apparent, not only in *War of the Worlds* but also in "The Crystal Egg" and "The Star," both of which include Martians. By this time he had also written three major novels: *The Time Machine, The Wonderful Visit,* and *The Island of Dr. Moreau;* in 1897 he would also pen *The Invisible Man.* All of these foreshadow in different ways elements that would appear in *War of the Worlds,* especially the evolutionary viewpoint. His son Anthony West believed that the seed of the idea is found in *The Time Machine,* where Wells spoke of humanity evolving into "something inhuman, unsympathetic, and overwhelmingly powerful."

All of this was in the background, but Wells himself tells us the circumstances that led him to set pen to paper in 1897: "The book was begotten by a remark of my brother Frank. We were walking together through some particularly peaceful Surrey scenery. 'Suppose some beings from another planet were to drop out of the sky suddenly,' said he, 'and begin laying about them here!' Perhaps we had been talking of the discovery of Tasmania by the Europeans – a very frightful disaster for the native Tasmanians! I forget. But that was the point of departure." We cannot say whether or not Wells's Martian novel would have been written without the offhand remark of his brother, but once the seed was planted, all the background we have detailed came into play.

The invention of the alien unfolded with the plot of the story, a story simple in concept but ingenious in execution, full of scientific ideas gleaned from evolutionary biology and astronomy. It is not just that Martians invade the Earth, blindly laying waste to all in their path. Wells invests his Martians with a certain plausibility; reason informs their every action, and their very presence betrays our smallness in an enormous universe.

Wells's "minds that are to our minds as ours are to those of the beasts that perish" were not a haphazard creation of his imagination. If the nebular hypothesis has any truth, Wells pointed out, Mars must be older than our world. Its smaller size by comparison with the Earth would have cooled it faster to a temperature where life could begin. It has a much more attenuated atmosphere, lower temperature, a surface only one-third ocean, and huge snow caps. The planet is in its last stages of exhaustion, and so its inhabitants look across space at the Earth, "a morning star of hope . . . green with vegetation," crowded only "with what they regard as inferior animals": creatures as alien and lowly as are the monkeys and lemurs to us. And so the plot is set, for "to carry warfare sunward is, indeed, their only escape from the destruction that generation after generation creeps upon them."

A central element of the story was triggered by an article Wells had seen in *Nature,* the British scientific journal, reporting the observation of a strange light seen on Mars. And so in the story, during the opposition of 1894 – the period during which Lowell began his scrutiny of the Martian surface – a light is seen on the surface of Mars, followed by an outbreak of incandescent gas, signs that Wells attributes to the "huge gun" by which the Martians later sent the spacecraft to earth. More signs were seen during the next two oppositions, and finally the spectroscope analyzed a mass of hydrogen gas moving toward the Earth with enormous velocity. The local astronomer scoffed at the very idea of Martians, but thought a meteorite shower was striking Mars or a volcanic explosion. Organic evolution had most probably not taken the same course on two adjacent planets, he argued – "the chances of anything man-like on Mars are a million to one."

One by one, during the course of the story, the missiles land on the Earth. They are believed at first to be meteorites, but eventually two creatures emerge from the first missile. As the reader continues, Wells's description is clearly meant to instill horror, for the tentacles that emerge first do not indicate something humanlike. And then, with the appropriate metaphor of the creatures emerging from the cylinder, we have the birth of the monstrous alien:

> A big greyish rounded bulk, the size, perhaps of a bear, was rising slowly and painfully out of the cylinder. As it bulged up and caught the light, it glistened like wet leather. Two large dark-coloured eyes were regarding me steadfastly. The mass that framed them, the head of the thing, it was rounded, and had, one might say, a face. There was a mouth under the eyes, the lip-less brim of which quivered and panted, and dropped saliva. The whole creature heaved and pulsated convulsively. A lank tentacular appendage gripped the edge of the cylinder, another swayed in the air.

This was only the first of Wells's many detailed descriptions of the alien – desperate, unfriendly, monstrous. That the alien appearance is more than skin deep Wells makes clear in his occasional descriptions of Martian physiology. The Martians are heads, with no digestive organs, and communicated by telepathy. They do not sleep, are sexless, and reproduce by budding, a direction in which Wells sees evolution might carry humanity. For mobility in the greater gravity of Earth, they construct machines. Ironically, the Martians are most like humans in their penchant for violence. When a deputation is sent with a flag to signal the good intentions of humanity, it is destroyed by the Martian heat ray. The Martians in their machines move toward London, wreaking destruction, and victory seems to be theirs. But in the end bacteria against which their bodies are defenseless destroy them, and the Martians lie slain "after all man's devices had failed, by the humblest things that God, in his wisdom, has put upon this earth."

The immense success of *The War of the Worlds* launched not only Wells's career, but also the career of the alien. Reprintings, variations on the theme, and imaginative art began immediately (Fig. 4.3) and have continued to the present. Adaptations have sometimes changed the setting, as in the 1938 Orson Welles American radio version. But always the effect has been powerful, and the imitations, variations, and elaborations on the story have echoed down the century. As late as 1908 Wells cited "the work of my friend, Mr. Percival Lowell," as testimony that *The War of the Worlds* was not too farfetched. Pronouncing Lowell's case for canals created by intelligent Martians "very convincing," Wells discussed the forms that such Martians might take. Although Lowell's claims were soon discredited, Mars remained a favorite setting for the alien throughout many decades. And after that, there were other settings around other stars; the possibilities were almost unlimited.

The invention of the alien, then, occurred independently in different forms in France, Germany, and England, in the latter two cases spurred on by Lowellian claims in the United States. But in all cases the evolutionary worldview and the plurality-of-worlds tradition played an important role in the birth of the alien. In France, Lamarckian and Darwinian evolution formed the background to the alien fiction of Flammarion, who had himself been France's chief champion of the plurality of worlds since 1862. The same background influenced Rosny, who undoubtedly knew Flammarion's work on plurality of worlds. In Britain a direct line exists from Darwin to Wells via Huxley amid the Proctorian plurality-of-worlds tradition. And in Germany, Lasswitz was influenced by the Kantian tradition. The alien thus emerged at the beginning of the 19th century from a confluence of science and philosophy, combined with the usefulness of the alien in exploring social as well as biological evolution.

4.2 THE DEVELOPMENT AND USES OF THE ALIEN: BURROUGHS TO BRADBURY

Once invented, the alien took on a life of its own. Though it might have died after the initial bursts of creativity of Flammarion, Rosny, Lasswitz, and Wells, once unleashed the alien proved a deep-seated and multipurpose concept. Its potential as a character in space adventure spanning the universe was obvious. More subtle were its uses in expanding areas of human thought like philosophy and religion, including such traditional problems as good and evil

Fig. 4.3. Cover from *Amazing Stories* of August 1927, illustrating H. G. Wells's *War of the Worlds*. Copyright 1927 by Experimenter Publishing Co.

and the uniqueness or triviality of humanity. And while such areas were being explored, the nature of the alien and interspecies communication were sub-themes of almost endless potential. Set adrift in the post-Darwinian era that had placed humanity at the pinnacle of terrestrial creation, the alien showed how provincial was the view of life on Earth, giving new meaning to Shake-speare's line "As flies to wanton boys are we to th' gods."

In this section we concentrate on the development and uses of the alien during the first half of the 20th century, the period bracketed by two ma-jor – but vastly different – American writers of alien literature, Edgar Rice Burroughs and the early Ray Bradbury. Although characterized by no sim-ple linear development, during this time the alien evolved from a rather pre-dictable character in the "space opera" adventures of Burroughs and others to the much more subtle, impressionistic, and almost ethereal creatures in Bradbury's *Martian Chronicles* (1950). The nature of alien morphology and behavior was also explored by several representative writers of the 1920s and 1930s. And the uses of the alien in philosophy and religion were especially evident in the work of two British writers: Olaf Stapledon and C. S. Lewis, who at the same time made their own contributions to defining the possibili-ties of alien morphology and psychology.

The decades immediately following its invention clearly demonstrate that alien literature did not need to be profound to be popular and influential. The alien could merely play the role of protagonist or antagonist in swash-buckling space adventures, without a serious attempt to illuminate its nature or the nature of humanity by comparison. That this lack of profundity did not translate into lack of popular impact (indeed, the relation may be inverse) may be seen in the work of Edgar Rice Burroughs (1875–1950), who took up Lowell's Martian theme in the United States several years before Lowell's death in 1916. It may well be true (as Arthur C. Clarke has said) that if one is no longer a teenager, it is too late to read Burroughs; it is no less true that for many during that impressionable age (including Bradbury, Clarke, and Carl Sagan), Burroughs's adventures under the moons of Barsoom (Mars) had a lasting effect on the imagination. Although perhaps best known as the inven-tor of Tarzan (1912), between 1912 and 1948 Burroughs also penned 10 vol-umes of the exploits of John Carter, battling or befriending various Martian life forms in the context of Martian culture. *A Princess of Mars,* published in novel form in 1917, was the first of the series, with *The Chessman of Mars* (1922) and *The Swords of Mars* (1936) considered among the best. Although Burroughs's Martians were no terrestrials – they had differently colored skin, variations in anatomy, and oviparous women, among other exotic character-istics – they were driven by essentially terrestrial passions. There is therefore little serious examination of either the alien nature or alien psychology and morality. All is subordinated to the story, which at times borders on fantasy.

None of this mattered to Burroughs's impressionable young fans. Action was the watchword of Burroughs's writing, and his readers loved it. None of the novelty of pure and simple space adventure of this type seemed to have worn off by the end of the century in similar stories such as *Star Wars*, also replete with aliens. Sophisticated or not, through such stories the alien gained a place in young minds, inspiring some of them to embrace careers in science or science fiction writing themselves. If nothing else, fanned by the imaginations of Lowell and then Burroughs, Mars was firmly entrenched in the United States as an adventurous and mysterious place in the solar system – a place one might want to visit some day.

Despite numerous examples of such adventurous alien "space operas" over the following decades, the alien phenomenon would hardly deserve serious attention if it had been used only for adventure. More serious questions abounded, among them the basic nature of the alien being. Lasswitz and Wells had already presented the two extreme possibilities – the extraterrestrial essentially good, though capable of retaliation, and the monstrous alien intent only on invasion. Although the invading monster carried the day throughout the 1930s, perhaps because of its dramatic story potential, a great deal of ground remained to be investigated between these two extremes. The works of three authors of the 1920s and 1930s demonstrate the possibilities inherent in the alien, both in physical characterization and in plot: David Lindsay's *A Voyage to Arcturus* (1920), Stanley G. Weinbaum's "A Martian Odyssey" (1934), and John W. Campbell's "Who Goes There?" (1938).

Unlike Lasswitz, Wells, and Burroughs, the Scottish-born writer David Lindsay (1878–1945) placed his aliens on the distant planet Tormance, orbiting Branchspell, the larger of two stars in the double star system known on Earth as Arcturus. But more than the setting separated Lindsay from his predecessors in alien literature; his influences were the music of Beethoven, the novels of a fellow Scottish author, George MacDonald, and the philosophical works of Schopenhauer and Nietzsche. *A Voyage to Arcturus*, the first and best of Lindsay's novels, went further than most in his imaginative characterization of alien lifeforms, and thus in giving the reader a feel for the truly alien. This is carried out in part by additional senses that Lindsay gives not only his aliens, but also the Earthling (Maskull), who upon landing on Tormance finds himself sprouting new organs that extend his perceptions of his new environment. With these continually developing new senses, Maskull interacts with the inhabitants of Tormance, whose skin is opalescent, its hue "continually changing with every thought and emotion," who communicate by telepathy, and whose habits (such as producing from the mouth a brightly colored crystal as an "overflowing of beauty") are completely foreign to terrestrials. The reader is subjected to a continually changing pageant of lifeforms, for life on the young planet is "energetic and lawless, and not sedate and imitative. Nature is still fluid – not yet rigid – and matter is plastic. The will forks and sports incessantly, and thus no two creatures are alike."

But these physical descriptions are only the setting for a plot that surpasses adventure and is at once philosophical, theological, and moral. The events surrounding Arcturus are not, anthropologist Loren Eiseley said in his Introduction to the 1963 edition of Lindsay's novel, simply "a superficial tale of odd beings with odd organs on a planet remote from our own. This is not a common story of adventure. Rather, it is a story of the most dangerous journey in the world, the journey into the self and beyond the self." Lindsay's journey is almost Dantean, but in an extraterrestrial setting, a modern-day exploration of self that nevertheless shows few traces of modern astronomy. Tormance was, of course, imagined, since no planets were known to exist around any star, but this did not matter; modern science was not Lindsay's concern. His concern was the search for a deeper beauty and reality than most people could commonly perceive, a reality he called Muspel. In encountering the many lifeforms in his journey across Tormance, Maskull is encountering beings who have, for different reasons, not yet found the deeper reality, a vision that he himself finally sees in the climax of *Arcturus*. By clearly using alien lifeforms to expand the sense of the possible and explore deep questions, Lindsay placed himself squarely in a nascent tradition of alien literature that went not only beyond Mars, but beyond invasion and adventure.

By contrast, much of the rest of alien literature seems tame, if not parochial, by the narrowness of its themes. But the continuing potential of new themes in the old Martian setting is clear in Weinbaum (1902–1935), who graduated with a degree in chemical engineering at the University of Wisconsin in the same class as Charles Lindbergh and made good use of his scientific background in his fiction. Though Weinbaum placed his first published science fiction story on Mars, his portrayal of the alien is generally considered a breakthrough: the Martians were not simply transported terrestrial monsters, but truly alien beings, intent on living their own lives rather than on invasion; the ideas of Flammarion and Rosny combined with the Martian myth to produce a totally alien world. Weinbaum's story featured an entire zoology of lifeforms: Tweel, the birdlike creature who befriends a visiting Earth expedition and humorously attempts communication; a silicon creature producing bricks as waste material; the "dream beast," a foreshadowing of Bradbury's work; and the people of the mud cities along the canals, more alien than any of the others. Here was a theme for the future: the study of alien mentality and motives and their reflection on the creature we call human. Reading such literature, one has the feeling that the horizons of life have been expanded, that the scope for new literary themes is endless, and that even the scientist might have his or her interest piqued by literary explorations extending beyond scientific claims.

As if to contrast the spectrum of alien possibilities yet again, 4 years after "A Martian Odyssey" appeared in *Wonder Stories,* John W. Campbell, Jr. (1910–1971), destined as both writer and editor to bring science fiction into the mainstream, used the alien as the central focus for a horror thriller published in *Astounding Science Fiction.* Having attended MIT and graduated from Duke University with a degree in physics in 1932, Campbell was another for whom science was not enough – it had to be supplemented by the imagination, and this he had done through science fiction beginning in 1930 with galactic epics incorporating his knowledge of science. "Who Goes There?", voted by the Science Fiction Writers of America as the greatest novella of all time, showed that the monstrous alien was not the sole preserve of unsophisticated writers (though there were many), and could in fact be used as the main character in a psychological thriller as great as any played out in a more mundane setting. Appearing in 1938, exactly 40 years after *The War of the Worlds* was published in novel form, Campbell's tale of a body-invading alien in the Antarctic amply demonstrated that original use of the monstrous alien had not been exhausted. Nor had it been exhausted 40 years later still, when the same theme was reworked in the movie *Alien.*

If Lindsay, Weinbaum, and Campbell extended our understanding of the physical possibilities of the alien and its uses, others would go beyond the physical to focus on its mental and moral capacities – as Lindsay had already shown was possible. No one did this with such flair and imagination as the British philosopher Olaf Stapledon (1886–1950). A graduate of Oxford in history and Liverpool University in philosophy (1925), Stapledon spent most of his life near Liverpool, though not in academic positions. He began writing essays as early as 1908, but only in 1930, at the age of 44, did he take up fiction, in which aliens immediately played a role. Though we meet many aliens in Stapledon's work, his aliens are not an end in themselves, nor is adventure the dominant theme. Rather, political, religious, and philosophical ideas predominate.

Three of Stapledon's works especially embrace the alien theme: *Last and First Men* (1930), *Star Maker* (1937), and *The Flames* (1947). The first covers a vast expanse of time – some 2 billion years – the second a vast expanse of both space and time. *Last and First Men* – which may show the influence of the biologist J. B. S. Haldane's *Possible Worlds* (1927) – tells the story of 18 species of humanity, some of them naturally evolved, others designed or bred, over a 2-billion-year span. In Stapledon's scenario, it was with the Second Men, a mere 10 million years in the future, that humans had their first encounter with extraterrestrial intelligence in the form of Martians. Like Wells 33 years earlier, Stapledon too had his Martian invasion, but it was only a moment in the span of the novel. Because Stapledon's Mars was also Lowell's dying planet, terrestrial water and vegetation were given as the Martians' rationale for invading Earth. But Stapledon's loftier reason for a Martian invasion was clear: it was intended to put humanity in perspective as a biological, spiritual, and philosophical entity. Millions of years later, driven from Earth by the approach of the Moon, Stapledon's Fifth Men escaped to Venus, a hot world almost completely covered by a shallow ocean. Here they found marine intelligence and other forms of life, "some sessile, others free-swimming, some microscopic, others as large as whales," all of them powered not by photosynthesis or chemical reactions, but by "the controlled disintegration of radioactive atoms." All of these lifeforms were reluctantly destroyed in the process of terraforming the planet for the needs of humanity. In due time Venus was abandoned for Neptune, and in the end it appeared that humans would be destroyed, except for one last attempt to seed their cells among the stars – an idea Stapledon undoubtedly took from the panspermia hypothesis. Thus, in the total expanse of 2 billion years of his story, extraterrestrials played a relatively minor role except for the Mars and Venus episodes.

Seven years later, in Stapledon's *Star Maker,* all the expanse of the previous novel was reduced to a few paragraphs, a "long human story, most passionate and tragic in the living . . . but an unimportant, a seemingly barren and negligible effort, lasting only for a few moments in the life of the galaxy." Unlike *Last and First Men, Star Maker* is replete with alien biologies that play an essential part in the story, but again the aliens themselves are not the main point. The purpose of the story is "not only to explore the depths of the physical universe, but to discover what part life and mind were actually playing among the stars." Ultimately, the aliens join the terrestrial protagonist in their search for the highest intelligence in the universe – Star Maker. And not only that: the joint purpose of the aliens and the terrestrial "was not merely scientific observation, but also the need to effect some kind of mental and spiritual traffic with other worlds, for mutual enrichment and community." The communication problem is solved by an enhanced form of telepathy in which the pilgrim "inhabits" the bodies of alien individuals.

In Stapledon's description of the different species of life and their search for Star Maker, the process of evolution, both biological and spiritual, was everywhere evident. In the latter regard, Stapledon made it clear that not all races had triumphed in what he called "a more awakened state"; for every one that did, hundreds of thousands ended in disaster. Stapledon indeed saw a great chain of being throughout the universe, with Star Maker at the apex. All the while, Stapledon did not forget his purpose: the search for the spirit of the universe, what humans on Earth formerly called God. Nor did he hesitate to draw poignant conclusions; midway in their journey, having seen the pain and suffering in the universe, the frequent destruction of civilization for every triumph, "it was becoming clear to us that if the cosmos had any lord at all, he was not that spirit but some other, whose purpose in creating the endless fountain of worlds was not fatherly toward the beings that he had made, but alien, inhuman, dark."

The immense span of time and space that Stapledon described was, in his view, only one creation in a vast series, an early mature work of Star Maker but one that, by comparison with later universes, was juvenile. "Musical" creatures with complex patterns and rhythms, universes with no time, worlds with antigravity rather than gravity, and reverse evolution were among Stapledon's concepts that provided an inexhaustible source of ideas to science fiction writers. In many ways Stapledon's work reflected Arnold Toynbee's just-published *A Study of History*, expanding the historian's discussion of civilizations beyond the Earth.

In the end, Stapledon succeeded in making his reader feel the smallness of humanity in the face of the vast universe, in which ultimate meaning was to be found. A measure of his success is found in the fact that it is difficult to overestimate the influence of his work, especially in Britain. J. B. S. Haldane, a prominent biologist in physiology and origins of life, knew Stapledon's work, as did C. S. Lewis, Arthur C. Clarke, and numerous others who went on to expound their own visions of meaning through literature. Like Stapledon, their visions of life in the universe served as vehicles for philosophical exploration, making use of contemporary knowledge of the physical universe, extrapolated to the biological universe in the search for a Cosmic Spirit.

If Stapledon's novels were a search for cosmic truth, the work of C. S. Lewis (1898–1963) was a defense of Christianity, a system that believed it had already found the truth, but for all that could still make use of the alien. Lewis, a literature professor at Oxford since 1925, would go on to achieve fame for his Christian apologetics targeted at both adults (*The Screwtape Letters*, 1942) and children (*The Chronicles of Narnia*, 1950–1956), while capping his career as a professor of medieval and renaissance literature at Cambridge University after 1954. At Oxford he was closely associated with his fellow Anglican, Charles Williams, and with the Roman Catholic fantasy writer J. R. R. Tolkien, both of whom composed works defending the Christian religion. It was in these surroundings that Lewis wrote a "Cosmic Trilogy" from 1938 to 1944, making heavy use of aliens and showing the influence of Stapledon's Mars and Venus episodes from *Last and First Men* and a more localized version of Star Maker. But it was Lindsay's *Voyage to Arcturus*, which Lewis read in 1936, that had the most immediate impact. Although he disagreed with Lindsay's philosophy, *Arcturus* was the springboard for what became the Cosmic Trilogy.

In *Out of the Silent Planet* (1938) a physicist and his colleague force a philologist to travel from Earth (Thulcandra) to Mars (Malacandra) to turn him over to the highest authority in return for gold. The philologist escapes, describes the many alien creatures he finds, and meets with the authority (Oyarsa) to discuss Earthly matters. The novel is a struggle between the spiritual and the scientific, probably based in part on the differences in worldview of Lewis (represented by the philologist) and J. B. S. Haldane, the Cambridge biologist, whose *Possible Worlds* (1927) also influenced Lewis. This work features a variety of Martians, in the form of sorns, hrossa, and pfifltriggi, and the spiritlike eldil, the latter with a physiology unlike that of humans or Martians. But it is the great eldil, Oyarsa, the ruler of Mars, who is most important to the theological story. Oyarsa "does not die . . . and he does not breed. . . . His body is not like ours, nor yours, it is hard to see and the light goes through it," one of the sorns explains. This creature tells the visitors that Earth is known as "The Silent Planet" for a reason: "Thulcandra [Earth] is the world we do not know. It alone is outside the heaven, and no message comes from it." Before life came to Earth, its Oyarsa became "bent," and he is forced to remain on Earth, where good must struggle with his evil. This enforced banishment prevents any taint of evil on other worlds. Through similar allegories, Lewis paints a vivid picture of the planet in which humans are in a spiritual struggle from which they must free themselves – precisely the same message conveyed in Lewis's nonfiction.

In *Perelandra* (1943) Lewis multiplied his aliens and continued his parable, this time on Venus (Perelandra), where a preparation for the struggle between good and evil takes place in *That Hideous Strength* (1944). Ever more anti-science, and in particular anti-Wells, *Perelandra* and *That Hideous Strength* drew the wrath of Haldane, among others. But taken as a whole, the novels received critical acclaim and a vast readership denied to the much more ambitious work of Lindsay, undoubtedly because of Lewis's more immediate appeal to Christian beliefs. All the while, even though the richly imagined aliens were in a sense only props for a story, the public interest in aliens was undoubtedly more than a little piqued. The years in which the trilogy was composed (1938–1944) span a crucial period of turnabout of belief in abundant planetary systems and thus in extraterrestrial life.

It is some measure of the continuing allure of Mars that by midcentury the planet that had inspired Burroughs's adventures of John Carter in 1912 (and indeed continued to inspire the last of his Martian novels in 1948) could still give birth to the more sophisticated aliens in the work of Ray Bradbury (1920–). It has been said of Bradbury that he "brought the traditional image of Mars to a kind of impressionistic perfection" in *The Martian Chronicles* (1950), a collection of linked stories that had appeared in the science fiction magazines of the late 1940s. It has also been said that Bradbury's ancestors are not Verne and Wells, conditioned by a scientific worldview, but Stapledon and Lord Dunsany. Incorporating elements of poetry and fantasy, *The Martian Chronicles* tells of human colonists who undertake a wave of expeditions to a still-canaled Mars, only to come literally face to face with their own past. Instead of Martians, they inexplicably find their hometowns populated by their own dead relatives brought back to life. But in reality they have found the Martians, who use not atomic weapons, but "telepathy, hypnosis, memory and imagination" to first lure and then kill the invaders from Earth. It is pure Bradbury that the "hard" science and technology of minutely described spaceships and weapons are absent, replaced with memory and imagination. Eventually the Earthlings win out, for chicken pox destroys the Martians' already dying civilization. Continuing waves of crass, greedy settlers proceed to despoil the planet Mars, as they had the Earth, in contrast to the "graceful, beautiful and philosophical people" who had created the Martian civilization. Finally, Earth destroys itself in an atomic war, and the Martian settlers rush back to Earth. Only a single family, their sensibilities about what is important in life renewed, returns to begin civilization anew on Mars.

Once again, the alien itself was not the main point of Bradbury's story, but rather a method for reflecting on what terrestrial civilization had become. The Martians had found a balance between religion, art, and science rather than letting science "crush the aesthetic and the beautiful." When the last family returns to Mars, the father tells his children, "Life on Earth never settled down to anything very good. Science ran too far ahead of us too quickly, and the people got lost in a mechanical wilderness, like children making over pretty things, gadgets, helicopters, rockets; emphasizing the wrong items, emphasizing machines instead of how to run the machines." Bradbury used science fiction to criticize science and technology rather than to glorify it – the diametric opposite of the goal of Wells and others. In this process the alien played an essential role, the detached and superior representative of life, even as it was dying, showing how civilization might develop along completely different principles.

If in the first half of the century science had come no closer to learning the truth about aliens, between Burroughs and Bradbury progress had surely been made in defining the possibilities. Readers became familiar with the alien as a character in space opera, but more than that, aliens had helped humans to explore traditional themes from a new and less parochial perspective. One sees in cosmic and theological alien literature a pattern of the search for a higher truth and wisdom, whether in Lindsay's Muspel or Stapledon's Star Maker. The character of the alien could range from the ridiculous to the sublime; perhaps because of that very flexibility, its career would accelerate in the second half of the century as the alien reached new heights of recognition in popular culture.

4.3 THE ALIEN COMES OF AGE: CLARKE TO E.T. AND BEYOND

Two developments characterize the coming of age of the alien in the second half of the 20th century. The further elaboration of old alien themes and the invention of new ones, of course, continued. But what was really new was the increasingly intimate relationship between science and science fiction, and the adaptation of these themes at an accelerating pace to the visually stunning and emotionally intense media of film and television. The visual media brought the alien one step closer to the hearts of the masses, while the marriage to science verified the alien theme as a plausible reality, not only because scientists themselves more frequently used fiction to speculate about alien contact, but also because alien science fiction influenced many who actually became involved in scientific programs to search for extraterrestrials. In other words, the role of the scientist as an author of alien literature in itself lent a credibility (though not verification) to the subject that it had not possessed in the first half of the century.

In the second half of the century the alien in literature became increasingly wedded to science in several ways: first, in the Stapledonian tradition of the cosmic perspective, as exemplified by Arthur C. Clarke; second, in the exploration of the nature of the alien in the tradition of the 1920s and 1930s pioneers, exemplified in the scientifically informed work of Hal Clement, Fred Hoyle, and Stanislaw Lem; and third, in the theme of contact with extraterrestrials, which now became more dominant as unidentified flying objects and radio searches for extraterrestrial intelligence brought the concept of contact to the fore. Even space opera was combined with serious scientific speculation about the relationship among intelligent species in the universe, as exemplified in the work of David Brin. The adaptation of some of these novels to visual media undoubtedly hastened science fiction's acceptance by the masses beyond the wildest dreams of its pioneers in the first half of the century. Perhaps more than purely scientific progress, these developments in science fiction go a long way toward explaining the pervasiveness of the belief in extraterrestrials in popular culture by the end of the 20th century.

The British author Arthur C. Clarke is prototypical of one whose preoccupation with the theme of alien encounter is used to place humanity in perspective while steeped in an adherence to the scientific tradition. At about the same time that Bradbury was discovering science fiction in the pulp magazines in America, Clarke (1917–) came to London as a civil servant in 1936. Here he worked as a radar officer in the Royal Air Force during World War II, predicted the development of communications satellites (1945), obtained a degree from Kings College (London) in physics and mathematics (1948), and served as president of the British Interplanetary Society (1946–1947, 1950–1953). His nonfiction books *Interplanetary Flight: An Introduction to Astronautics* (1950) and *The Exploration of Space* (1951) made him well known as an advocate of space travel long before he was known for his science fiction; although already in the 1930s he had begun work on several versions of what would become his first novel, *Against the Fall of Night* did not appear until

1953. The most important influence on this work was none other than Olaf Stapledon's *Last and First Men,* which Clarke discovered in his local library shortly after its publication: "With its multimillion-year vistas, and its roll call of great but doomed civilizations, the book produced an overwhelming impact upon me," Clarke recalled years later.

Like Stapledon, his British successor thought in terms of eons. His first novel, expanded in 1956 as *The City and the Stars,* described the far future on Earth in which technology had kept the utopian city of Diaspar running smoothly but completely isolated from the rest of the universe for a billion years. The story's protagonist, considering his civilization in a state of stagnation rather than utopia, travels across the galaxy, finds an alien intelligence, and initiates a cultural renaissance. Although most of the novel is spent with the protagonist (with the unpretentious name of Alvin) describing the city and trying to determine how to leave it, his first glimpse of the stars is an epiphany that leads on in the final quarter of the novel to the stars. There he finds humanity's past – a past in which humans had explored the Galaxy. In that age of exploration "everywhere he found cultures he could understand but could not match, and here and there he encountered minds which would soon have passed altogether beyond his comprehension." Returning to Earth to brood about their fate, humans began a great experiment of mind and genetics, spanning millions of years and involving other galactic civilizations, to build a pure mentality, a disembodied intelligence that would search for "a true picture of the Universe" unencumbered by physical limitations. This it did, at first resulting in disaster in a destructive Mad Mind but eventually culminating in a pure mentality that possessed knowledge but not wisdom. This entity was left to roam the universe while humanity, desiring peace and stability, retreated to Diaspar. It was this pure mentality that Alvin discovered, and from which the knowledge and inspiration are gained to once again strike out into the mysteries of the universe. Unlike Lasswitz, whose Earth found utopia only by keeping the Martians at a distance, Clarke found it only by seeking extraterrestrial civilizations, though in a much broader perspective than our own solar system. This search was to set a pattern in much of Clarke's science fiction, for Clarke, like his character Alvin, was an explorer.

Meanwhile, after a series of short stories dealing with the alien theme, in 1953 Clarke published *Childhood's End,* which quickly became a classic of alien literature. This time the Earth itself is visited by alien "Overlords" (Fig. 4.4) possessing a bodily form resembling the devil, who attempt to incorporate humanity into the scheme of universal sentience. The story is set in the immediate future as humans are about to journey to the Moon. The Overlords are benevolent aliens who impose an end to war and world government on Earth. Amid some resistance, the first contact theme is played out: the eradication of poverty and ignorance is set against negative aspects that

Overlord

Fig. 4.4. An Overlord from Arthur C. Clarke's *Childhood's End* (1953). The Overlords, agents of an even more superior intelligence, impose world government on Earth and usher humanity into a universal mind. Art from Wayne Douglas Barlowe from *Barlowe's Guide to Extraterrestrials,* copyright © 1979 by Wayne Douglas Barlowe. Used by permission of Workman Publishing Co., Inc., New York. All rights reserved.

include a decline in the creative arts and in science brought on by the Overlords' vastly superior knowledge. Yet the Overlords, they reveal later, are only agents of an even more superior intelligence, and their ultimate goal is to usher the human race (by way of its children) into the Overmind, something the Overlords themselves cannot achieve. As in *The City and the Stars,* the concern of the novel is with the ultimate destiny of humanity, which, though different in detail, still is intimately connected to extraterrestrials. Both novels involve a religious vision: in *City,* the creation of the disembodied intelligence was recognized as "a conception common among many of Earth's ancient religious faiths," and it was noted as strange that an idea that "had no rational origin should finally become one of the greatest goals of science"; in *Childhood's End* the children give up their individual souls to unite with the universal soul that is the Overmind. Fifteen years after *Childhood's End*, Clarke would still be seeking an extraterrestrial destiny, via yet another route, in the mystical denouement of *2001: A Space Odyssey*.

Driven by a belief that extraterrestrials gave a true perspective on humanity, Clarke made the alien a pervasive theme throughout his career. "The idea that *we* are the only intelligent creatures in a cosmos of a hundred million galaxies is so preposterous that there are very few astronomers today who would take it seriously," he wrote in 1972. "It is safest to assume, therefore, that *They* are out there and to consider the manner in which this fact may impinge upon human society." Although Clarke treated many themes in his long career, this real-life conviction about extraterrestrials was the dominant thread running through his science fiction. By the early 1950s, this theme was not only an echo from Stapledon, but also an idea reinforced by science with its claims for abundant planetary systems.

While Clarke excelled in the Stapledonian search for meaning in the cosmic context, others explored the potential strangeness of the alien in ways that went beyond Lindsay, Weinbaum, and their contemporaries in the first half of the century. In the United States, Hal Clement (1922–), who obtained degrees in astronomy, chemistry, and education and worked as a high school teacher, set his highly original novel *Mission of Gravity* (1954) on a rapidly spinning, discus-shaped planet (Mesklin), whose gravity varied from 3 times (at the equator) to 700 times (at the pole) that of Earth. The quintessential "hard" science fiction writer, Clement elaborated his story around the natives who, having necessarily adapted to their world, were only 15 inches long and 2 inches in diameter and operated on a vastly different time scale than humans.

Clarke's British contemporary, the eminent astronomer Fred Hoyle (1915–), explored a wholly different kind of alien than Clement's, though with similarities to the disembodied intelligence of Clarke's *The City and the Stars*. Known as a brilliant maverick in the astronomical community, Hoyle was among the first astronomers to champion abundant solar systems and the implied possibilities of life. This was undoubtedly the background to his first novel, *The Black Cloud* (1957), which demonstrates that while science could inform fiction, it need not keep a tether on imagination. The alien intelligence in this book takes on the form of a cloud of interstellar matter, some half-billion years old, with which astronomers are eventually able to communicate. The cloud, which initially causes chaos and threatens destruction of the Earth, is convinced to retreat, showing the potentially benevolent character of intelligence. But not before Hoyle has the chance to explore the nature of both intelligence and communication. Planetary intelligence, the cloud reveals, is unusual because the gravitational force limits the size of its beings and the scope of their neurological activity, and the comparative lack of chemical food leads to the "tooth and claw existence" characteristic of Darwinian survival of the fittest. When radio communication is established with the cloud, the detailed comparisons of neurological structure between an intelligent cloud and a human are discussed, giving the sense of just how unique – and inferior – humanity might be. Significantly, Hoyle was a scientist for whom extraterrestrials played an important role in his worldview; he later wrote much more on extraterrestrial life, both in fiction and nonfiction.

An even stranger exploration of alien nature emerged from the non–Anglo-Saxon world in the form of Stanislaw Lem's *Solaris* (1961), in which humans attempt to communicate with a living ocean on the planet Solaris. Lem (1921–) was a Polish physician who wrote on the history and methodology of science (particularly cybernetics) and ran into trouble with Soviet Lysenkoism. He began writing science fiction in about 1950, as Bradbury's *Martian Chronicles* was being published in America. Although Lem was not affected by the Martian mania in the West, he was concerned with many of the same problems of human identity and purpose, and he made use of the human mind to conjure living beings in a way reminiscent of Bradbury. But in *Solaris* he was able to play out these concerns in an alien setting unlike anything produced in the West. When the planet Solaris is found to contain an ocean that is in some sense alive, the space station Solaris hovers a few hundred meters above the surface of the ocean for the purpose of examination. Countless attempts in the past to establish contact had been frustrated; now much was learned about the ocean, but in the end scientists were convinced only "that they were confronted with a monstrous entity endowed with reason, a protoplasmic ocean-brain enveloping the entire planet and idling its time away in extravagant theoretical cognitation [*sic*] about the nature of the universe. Our instruments had intercepted minute random fragments of a prodigious and everlasting monologue unfolding in the depths of this colossal brain, which was inevitably beyond our understanding."

Communication of a sort finally emerges from the scientists' subconscious minds as the ocean synthesizes out of each scientist's past a "Phantom" living person constructed of neutrinos. There are parables here about human communication and the meaning of life, but with regard to alien life one important question raised (very rarely in other science fiction) is whether the very nature of the alien will prevent human communication with it: Lem portrays the attempts at contact "like wandering about in a library where all the books are written in an indecipherable language. The only thing that's familiar is the color of the bindings!" Beyond the issue of comprehensibility, there is the question (also submerged in most science fiction) of whether contact *should* be made. "We think of ourselves as the Knights of the Holy Contact," Lem has one of his characters say. "This is another lie. We are only seeking Man. We have no need of other worlds. . . . We are searching for an ideal image of our own world: we go in quest of a planet, of a civilization superior to our own but developed on the basis of a prototype of our primeval past." And again, "Solaristics is the space era's equivalent of religion: faith disguised as science. Contact, the stated aim of Solaristics, is no less vague and obscure than the communion of the saints, or the second coming of the Messiah. Exploration is a liturgy using the language of methodology; the drudgery of the Solarists is carried out only in the expectation of fulfillment, of an Annunciation, for there are not and cannot be any bridges between Solaris and Earth." While the ultimate purpose of Lem's novel is to use the cosmos to learn about humans, it may also be read at a different level as an argument against attempting contact before humans understand themselves: "Man has gone out to explore other worlds and other civilizations without having explored his own labyrinth of dark passages and secret chambers, and without finding what lies behind doorways that he himself has sealed." Unlike the search for meaning of Lindsay, Stapledon, and Clarke, Lem's search failed, imparting the message that our fate may lie not in the stars, but in ourselves. Lem's novels, including *Solaris* (the first to be translated into English), were widely read in both the East and West, and *Solaris* was filmed in 1971.

Despite Lem's warnings of alien incomprehensibility and human infancy, numerous science fiction novels were inspired by real-life efforts to detect extraterrestrial intelligence by radio telescopes, beginning with Frank Drake's Project Ozma in 1960. Although already explored in fiction by Hoyle's *A For Andromeda* (1962) and Harrison Brown's *The Cassiopeia Affair* (1968), the theme found perhaps its classic expression in James Gunn's *The Listeners* (1972). In Gunn's novel the frustrations of a long search are vividly portrayed against a backdrop of "computer runs" that relate the history of the idea of other worlds. Gunn treats not only the institutional workings and frustrations, but also the possible implications of contact, with his portrayal of Jeremiah, the leader of a religious cult that denies the existence of extraterrestrial intelligence.

All of these themes – cosmic perspective in the context of extraterrestrials, original explorations of the nature of the alien, and contact with extraterrestrial intelligence – would continue to be elaborated in ever more subtle form throughout the century. In the tradition of the cosmic perspective, subsequent authors did not often reach the metaphysical heights of Stapledon and Clarke (except for Clarke himself), but there was ample scope for examining narrower themes such as C. S. Lewis's focus on Christianity. In this line James Blish's *A Case of Conscience* (1958) portrayed a Jesuit priest and biologist who must deal with the implications of the discovery of intelligent reptilian inhabitants on the planet Lithia. Imaginative portrayals of the alien in the tradition of Clement also continued, notably with Robert Forward, whose *Dragon's Egg* (1980) and *Starquake!* (1985) described life on a neutron star, an extremely dense object where a generation passes in 37 minutes. David Brin intelligently explored even the idea of life on the Sun in *Sundiver* (1980). And the idea of radio contact was explored further by one of the pioneers in the field, Carl Sagan, whose novel *Contact* (1985) is grounded in a scientific search that discovers, as in Arthur C. Clarke's novels, that humanity's true destiny is among the stars.

At the same time, a more sophisticated space opera in the vein of Burroughs has never lost its charm or its readership. In this tradition is David Brin's *Startide Rising* (1983) and *The Uplift War* (1987). The continued acceptance of this tradition in the field of science fiction is evidenced by the fact that these novels captured the field's highest honor, the Hugo Award.

Although we have seen only the tip of the alien iceberg in this examination of some of the classics of alien literature, it is equally important to note that the alien was not a major theme for every writer of science fiction. The most notable example in the United States is Isaac Asimov (1920–1992). Though Asimov is often compared to Arthur C. Clarke for popularity and other reasons, the contrast in their attitude toward extraterrestrials is striking: the universe of Asimov's Foundation series and that of his robot series (eventually related to each other) are both devoid of aliens. Similarly, though Asimov had featured aliens in a few of his early short stories, his other mainstream novels also lack aliens. Only a relatively late work, *The Gods Themselves* (1972), is notable for its alien beings, but the result, in one critic's words, was "among the most fascinating and believable aliens yet imagined in science fiction." Still, it is important to realize that for even some of the most widely read science fiction authors, extraterrestrials were not an essential part of their worldview or a criterion for public acceptance.

Moreover, it is well to remember that the alien theme was, of course, only a small part of the total scope of science fiction. Themes of the relation of humanity to the universe could be carried out without the alien, as could the other themes discussed earlier. Furthermore, one could argue that the "New Wave" science fiction beginning in the 1960s, by eschewing traditional themes, also tended to avoid the well-worn subject of the alien more than did conventional science fiction. Nevertheless, the amount and scope of alien literature in the 20th century are truly impressive, as is its appeal to the public. Nowhere is this appeal more evident than in the success of science fiction in the visual arts, a success that seemed to have no bounds in the second half of the century.

For all of its imagination and innovation, science fiction – and the alien with it – would have remained the province of a limited readership had it not been for the expansion of the genre to the visual media. The alien boom in cinema had begun almost exactly at the turn of the half-century, for reasons that probably have little to do with science and more to do with the onset of the Cold War and the paranoia about Communists. Whatever the reason, cinema seems to some extent to have repeated the evolution of the alien in science fiction literature, from the unsophisticated aliens of the early 1950s to the maturity of the mysterious, existential, and yet unseen aliens in Arthur C. Clarke and Stanley Kubrick's *2001: A Space Odyssey* and the technologically superior aliens in the movie version of Carl Sagan's *Contact* (1997).

The alien boom in cinema is generally considered to have begun with *The Thing* (1951), based on John W. Campbell's "Who Goes There?" (discussed earlier). Called "a superior example of sf cinema" and "by far the most influential of the films that sparked off the sf/monster movie boom of the 1950s," it featured James Arness as a humanoid vegetable who arrives on an unidentified flying object (UFO) and terrorizes an Arctic base. In the same year appeared the morally more sophisticated film *The Day the Earth Stood Still*, in which the alien Klaatu and his robot attempt to stem human violence. The year 1953 saw several notable alien films, including a vastly altered *War of the Worlds*, now set in contemporary California instead of 1890s England. *It Came from Outer Space* (1953), based on "The Meteor" by Ray Bradbury, portrayed aliens whose spaceship crash lands, who are able to change shape, and who duplicate humans to assist in the repair of their spaceship. *This Island Earth* (1955) featured a scientist who is taken to another planet in the hope that he can help keep its atomic shield functioning against attacks from enemies – a clear analogy to the Cold War in full swing on Earth. Although not generally considered a great film, it has been called "an excellent bad film." With *Invasion of the Body Snatchers* (1956), which has been labeled "a subtle and sophisticated movie . . . possibly the most discussed B-grade movie in the history of American film"; *Earth vs the Flying Saucers*, which capitalized on the UFO mania spreading across the world and climaxed with a battle over Washington, D.C.; and, in the same year, *Forbidden Planet*, with its aliens both alive and vanished, the alien was firmly entrenched in the cinema. The dawn of the Space Age the following year could only heighten the interest – and the potential realism – of such movies.

The 1960s saw the alien in film begin to mature in quite remarkable fashion. While alien films continued, the *Star Trek* television series, which ran initially from 1966 to 1969, became something of a popular culture phenomenon. *Star Trek*, which reminded its viewers at the beginning of each episode that its mission was "to seek out new life and new civilizations," carried out many of its alien episodes in imaginative and memorable fashion. Created by Gene Roddenberry, the television series was so popular that it spawned a following of "Trekkies," some of whom were undoubtedly fascinated with space travel in general, but others of whom were much taken with the treatment of the alien theme. It also spawned a new television series, *Star Trek: The Next Generation* (1987), eight feature films from 1979 to 1996, and well-attended Star Trek conventions.

Perhaps even more significant for the alien theme during the tumultuous decade of the 1960s, which witnessed assassinations on the one hand and the Apollo Moon landing on the other, Arthur C. Clarke's vision of cosmic perspective reached the cinema with *2001: A Space Odyssey* (1968), making spectacular but restrained use of the alien theme. It was based on his short story "The Sentinel," written in 1948. His mystical treatment, in the hands of producer/director Stanley Kubrick, raised science fiction film to a new level. Five years and $10 million in the making, it produced mixed reviews but quickly became a classic discussed by many, if not understood by all. It is significant that what made *2001* so popular and influential was not the book, but the movie; only after the success of the movie did Clarke become the most popular science fiction writer in the world. Although the sequel novels and film, *2010: Odyssey Two* (book, 1982; film, 1984), *2061: Odyssey Three* (book, 1988), and *3001: The Final Odyssey* (book, 1997), were less successful, the impact of *2001* itself would be felt throughout the century.

During the 1970s two spectacularly successful movies featured aliens. In *Star Wars* (1977) the variegated aliens were part of space adventure, expanded from Burroughs's Mars to galactic space, as others had often done before in science fiction but that few novels had accomplished on film. As the freedom fighters tackled the evil Galactic Empire, aliens played a less than cerebral role. The continuing popular success of its sequels, *The Empire Strikes Back* (1980) and *The Return of the Jedi* (1983), with their renderings of Yoda, Ewoks, and Jabba the Hutt, demonstrate that space opera was no less popular at the end of the century than when Burroughs invented the genre.

The other film, *Close Encounters of the Third Kind* (1977), played on a more down-to-Earth subject: the continuing fascination of the public (if not scientists) with UFOs. Directed by Steven Spielberg, the movie took its name from the categorization of UFOs by astronomer J. Allen Hynek (who made a cameo appearance), the "third kind" being physical contact (see our next chapter). The movie vividly evoked both the anxiety and the sense of wonder that extraterrestrials can generate, culminating in the final scene in which humans and aliens communicate by means of musical tones.

Finally, to come full circle from our discussion of the works of Lasswitz and Wells in 1897, the last quarter of the century saw cinema explore the two opposite extremes of alien nature: the good and the evil. *Alien* (1979), followed by *Aliens* (1986), *Alien III* (1992), and *Alien Resurrection* (1997), evoked a vivid horror as the human visitors to a planet are attacked by creatures, one of which reproduces inside a crewman's body and escapes to terrorize the crew.

The wildly popular *Indpendence Day* (1996) once again graphically depicted aliens intent on invading Earth. By stark contrast, *E.T.: The Extraterrestrial* (1982) depicted a lovable alien left behind when his spacecraft hurriedly departed. Befriended by children even as he is hunted by suspicious government scientists, E.T. learns some of Earth's customs through his friends but becomes ill, dies, is resurrected, and with the help of the children escapes to the spaceship that returns for him. The fact that *E.T.* became the most commercially successful science fiction movie of the century until *Jurassic Park,* a decade later, demonstrated that the public preferred benevolent aliens to horrific ones; the *Alien* series, although popular, did not approach the success of *E.T.* Popularity, however, had nothing to do with reality, and the fact remained that while *E.T.* and *Alien* starkly portrayed alternative alien natures, scientists had no idea which form might actually predominate.

By the end of the century the alien, barely invented 100 years before, had come to assume a central role in popular culture and scientific imagination. We may trace, during the course of that century, several families of alien literature drawing upon a variety of traditions. The monstrous alien, beginning with Wells, extending through John W. Campbell and on to *Aliens,* is certainly one of the strongest and most vividly portrayed in film. The swashbuckling alien adventures prototyped by Burroughs extend through numerous authors in a direct line to *Star Wars.* More difficult to portray in film was the use of the alien in cosmic perspective, from the spiritual journey of Lindsay to the vast expanse of space and time in Stapledon, and thence from Lewis's limited Christian journey to Clarke's more cosmic vision; nevertheless, *2001: A Space Odyssey* succeeded in capturing on film even this alien genre. Finally, the hope for utopia via extraterrestrials, so optimistically treated in Lasswitz at the end of the 19th century, still found its adherents in authors like Clarke, a product of the British empire, but also drew its pessimists in authors like Lem, who had experienced firsthand the result of the Russian empire.

During the century, science and science fiction increasingly complemented each other: speculative science fiction provided the perfect outlet for scientists who wished to go beyond science. Not only did scientists exercise their imaginations in science fiction, science fiction also inspired them to tackle questions in the real world. Many of the pioneers in exobiology and the Search for Extraterrestrial Intelligence (SETI) grew up on science fiction and were led to their careers by its imaginative lure. Having nurtured science fiction, science now received in return some of the rewards of imagination. This fact alone gives science fiction a prominent place in the 20th century's fascination with extraterrestrial life.

A final truth must also be recognized. Despite all the advances of the 20th century in uncovering the secrets of astronomy, terrestrial biology, and evolution, science could as yet add nothing to the question of the physical, mental, and moral nature of intelligence beyond the Earth. At best science might shed a pale light on the question of the possible physical forms of the alien, but it could say nothing about its mental evolution – much less whether good or evil or some compromise between the two ruled such intelligences as might exist in the universe. For that, the speculations of science fiction – from the monstrous form conjured in the work of Wells, Campbell, and *Aliens* to the enlightened beings of Lasswitz, Lindsay, and *E.T.* – were as valid as anything science could suggest. With all its science, this most crucial question of all remained for the 20th century in the realm of the imagination.

5

THE UFO CONTROVERSY AND THE EXTRATERRESTRIAL HYPOTHESIS

Horatio: O day and night, but this is wondrous strange!
Hamlet: And therefore as a stranger give it welcome.
There are more things in heaven and earth, Horatio,
Than are dreamt of in your philosophy.

Shakespeare, *Hamlet*

What we have here is a signal-to-noise ratio problem: There is indeed a fantastic amount of noise, represented by the many misidentifications of familiar objects seen under unusual or surprising circumstances – balloons, birds, satellites, meteors, aircraft, stars – yet, in all scientific honesty, one is led to ask whether there might not indeed be a signal somewhere in the noise.

J. Allen Hynek (1966)

This comprehensive further examination of the so-called "Roswell Incident" found no evidence whatsoever of flying saucers, space aliens, or sinister government coverups.

Headquarters, U.S. Air Force
The Roswell Report: Case Closed (1997)

How often have I said to you that when you have eliminated the impossible, whatever remains, however improbable, must be the truth.

Donald Menzel quoting Sherlock Holmes (1953)

No one would have believed in the last years of the 19th century, even as H. G. Wells was writing his *War of the Worlds,* that the 20th century would witness on a sporadic but large scale a strange new phenomenon in the skies of planet Earth; that this phenomenon, eventually known as "flying saucers" or "unidentified flying objects (UFOs)," would become the subject of raging controversy among scientists and the public alike; and that – for better or worse – this controversy would become intimately associated with the debate over the existence of extraterrestrial intelligence. That this is precisely what happened is surely one of the most remarkable tales of the 20th century. More remarkably still, as one of the principal scientists involved would point out, the UFO controversy was unlike other novel scientific ideas such as evolution, in which scientists tried to impose new ideas on the masses. Rather, like the 19th-century report of stones falling from the sky (now known as meteorites), it was the masses who tried to impose a novel idea on a mostly incredulous community of scientists. Between public gullibility and scientific closed-mindedness, between perception and reality, lies an important chapter in the history of the extraterrestrial life debate and a story of the limits of science under the most trying circumstances.

137

UFOs are the subject of an enormous literature, some of it of more interest to the social sciences of psychology and sociology than to the physical sciences. It is not our purpose to review this literature or to analyze whether UFOs really exist. Nor is it our intention to write the history of the UFO debate. Rather, we wish to examine the interaction of the UFO controversy with the extraterrestrial life debate. This immediately narrows our focus to what historically is known as the "extraterrestrial hypothesis" of UFOs – the idea that this mysterious aerial phenomenon represents spaceships from alien civilizations. Even here, we are not interested for present purposes so much in the details of specific cases as in the scientific and public reaction to the extraterrestrial hypothesis, how this reaction reflects on the scientific and public attitudes toward the existence of life beyond the Earth, and the nature of the arguments scientists brought to bear in reaching their conclusions. Although UFOs have been a worldwide phenomenon, for practical purposes of available documentation we focus chiefly on the United States, which has been the major center of UFO activity and study. In this chapter, then, we examine the rise of the extraterrestrial hypothesis beginning in 1947, how it peaked in scientific attention in the late 1960s with activity centered on the U.S. Air Force–sponsored "Condon Report," and the subsequent decline in the scientific community of extraterrestrials as a serious hypothesis for the elusive objects seen in the sky.

5.1 THE RISE OF THE EXTRATERRESTRIAL HYPOTHESIS

The rise of the extraterrestrial hypothesis of UFOs to 1965 is characterized by a central role for the media, a schizophrenic attitude by the U.S. Air Force, and (with two outstanding exceptions) the lack of substantial scientific participation. While much of the media – with their usual mixed motives of truth and profit – continually pushed the extraterrestrial hypothesis on a receptive public, and while the Air Force was understandably preoccupied with national security aspects of the issue, scientists in many ways abdicated their role as critical analyzers of an unexplained phenomenon. In part this was due to the reluctance of scientists to engage in a controversial issue that would do nothing to advance their careers and might do their reputations great harm. In part the fleeting nature of the phenomenon and the resulting lack of hard data precluded easy scientific analysis, at least by way of the normal scientific methodology. And perhaps most significant, the concept of extraterrestrial intelligence – the favored explanation for UFOs among the media and the public – was not yet a part of the collective scientific consciousness, as it certainly would be in later years. For all these reasons, the UFO controversy to 1965 assumes the surrealistic character of mostly nonscientific individuals and organizations thrashing about in the midst of a potentially significant

Table 5.1. *Number of UFO reports received each month by Project Blue Book, 1950–1968*[a]

	J	F	M	A	M	J	J	A	S	O	N	D	Total
1950	15	13	41	17	8	9	21	21	19	17	14	15	210
1951	25	18	13	6	5	6	10	18	16	24	16	12	169
1952	15	17	23	82	79	148	536	326	124	61	50	42	1501
1953	67	91	70	24	25	32	41	35	22	37	35	29	509
1954	36	20	34	34	34	51	60	43	48	51	46	30	487
1955	30	34	41	33	54	48	63	68	57	55	32	25	545
1956	43	46	44	39	46	43	72	123	71	53	56	34	670
1957	27	29	39	39	38	35	70	70	59	103	361	136	1006
1958	61	41	47	57	40	36	63	84	65	53	33	37	627
1959	34	33	34	26	29	34	40	37	40	47	26	10	390
1960	23	23	25	39	40	44	59	60	106	54	33	51	557
1961	47	61	49	31	60	45	71	63	62	41	40	21	591
1962	26	24	21	48	44	36	65	52	57	44	34	23	474
1963	17	17	30	26	23	64	43	52	43	39	22	22	399
1964	19	26	20	43	83	42	110	85	41	26	51	15	562
1965	45	35	43	36	41	33	135	262	104	70	55	28	887
1966	38	18	158	143	99	92	93	104	67	126	82	40	1060
1967	81	115	165	112	63	77	75	44	69	58	54	24	937
1968	18	20	38	34	12	25	52	41	29				

[a] Sum of reports received from Air Force bases and those received directly from the public.
Source: E. U. Condon, *Scientific Study of Unidentified Flying Objects* (New York, 1969), Section V, Table I.

phenomenon, resulting in sporadic reaction rather than systematic study. And yet, for all that, it is unique in eliciting public and scientific attitudes toward extraterrestrial life that otherwise would never have been expressed.

One of the peculiarities of the UFO phenomenon is its wavelike character in terms of numbers of sightings, at least in the United States. As documented by the historian David Jacobs in *The UFO Controversy in America,* the first wave of "mystery airships" came in 1896–1897 – before the Wright Brothers succeeded in heavier-than-air flight – followed by a half-century lull until the beginning of the modern debate in 1947 and then succeeding waves in 1952, 1965–1967, and 1973. This wavelike character of the phenomenon in the United States, against the steady background of yearly reports, may be seen in Table 5.1 for the years 1950–1968. In what follows, we shall be alert to any correlation of these peaks with concurrent developments, especially in science.

It is notable that the idea that the UFO phenomenon might be due to extraterrestrial spaceships was mentioned even in the first wave. The 1896–1897 sightings, after all, appeared to be airships, and although the favored explanation was secret experimental aircraft controlled by individuals or governments, the other alternative was that the supposed aircraft came from beyond the Earth. The few astronomers who offered an opinion agreed for the most part that there was a more mundane astronomical explanation, such as the planet Venus or a bright star; the idea of extraterrestrial spaceships, possibly from Mars, came from the popular side rather than the scientific. Even Percival Lowell, who had originated his theory of Martian canals in 1894 and was just beginning his long battle for acceptance of that idea, does not appear to have offered an extraterrestrial hypothesis for UFOs. But that did not prevent the press from doing so, probably under the influence of Lowell's writings. In 1897 the *St. Louis Post Dispatch,* stating that "these may be visitors from Mars, fearful, at the last, of invading the planet they have been seeking," suggested that a message of peace and an invitation to land be sent, though it did not say how. Both the *Houston Post* and the *Washington Times* suggested links of the phenomenon to visitors from the planet Mars, while a Tennessee paper rejected the extraterrestrial hypothesis on the grounds that anything like human life could not be sustained during the voyage. But like the 1896–1897 wave itself, the extraterrestrial hypothesis for UFOs faded into history for a half-century. Although the reasons for this have never been analyzed, it is remarkable that while the idea of Martian intelligence rose to full fury until 1910, the reports of UFO-like phenomena were confined to a 2-year period at the end of the 19th century. This clue deserves more attention in any study of the UFO phenomenon. Lowell's ideas of Martians offered the perfect scientific excuse for a continued UFO wave, and yet the public, while enamored of Lowell's vision, did not transport the Martians to terrestrial skies – at least not until the 1938 Orson Welles broadcast. And Lowell's silence indicates that even he placed limits on what was considered good scientific evidence for extraterrestrials.

Whatever the reason, the modern era of UFOs began only in 1947, and it was not long before the extraterrestrial hypothesis was put forth as a possible explanation. This time it would be the opening shot of a sustained hypothesis. Although mysterious sightings had been sporadically reported earlier, on June 24, 1947, Kenneth Arnold, flying his private plane near Mt. Ranier in Washington State, reported nine disc-shaped objects flying in formation at speeds he estimated to be at over 1000 miles per hour. Arnold, a reputable businessman and deputy U.S. marshal, was taken seriously, and his description of the objects as flying "like a saucer if you skipped it across the water" led the newspapers to coin the term "flying saucer." His report precipitated more than 850 sightings during the year from others. A variety of explanations were offered, but according to a Gallup poll at the time, very few people immediately sought an extraterrestrial explanation. Although within months 90 percent of the population had heard of flying saucers, most thought they were illusions, hoaxes, secret weapons, or other earthly phenomena. Once again, it was not the public that seemed anxious to court extraterrestrials, and although astronomers were just beginning to hint that life might be common among the distant stars, they were certainly not prepared to associate this idea with a UFO phenomenon that implied that such intelligence had actually reached the Earth.

The media, on the other hand, were more than willing to exploit extraterrestrials. In particular, Arnold's book *The Coming of the Saucers* (1952) was coauthored and published by Ray Palmer (1910–1977), the managing editor of the science fiction magazine *Amazing Stories,* who was promoting the extraterrestrial hypothesis even before the Arnold sightings. Although Palmer's magazine reached only a specialized group, his suspicion of government and his inability to separate science fiction from science fact are characteristics that would mark the extraterrestrial hypothesis of UFOs for many years. The increasing interest in science fiction, which heavily emphasized extraterrestrials (Fig. 5.1), should not be underestimated as a contributor in itself to the rise of the extraterrestrial hypothesis.

The media and science fiction magazines were one thing, but ironically in view of subsequent history, it was only when the U.S. Air Force decided to investigate the flying saucer reports that the extraterrestrial hypothesis was recognized at an official level. During 1947 the Air Force, charged with the security of the skies for the United States, collected 147 flying saucer reports at its Technical Intelligence Division of the Air Materiel Command at Wright Field in Dayton, Ohio. On December 30, 1947, the order was given to begin a project to study the phenomenon, and an incident on January 7, 1948, reinforced the propriety of Air Force participation. On that date a large number of people spotted a UFO in proximity to Godman Air Force Base, near Louisville and Fort Knox, Kentucky. When three F-51 planes, led by Captain Thomas Mantell, went to check out the reports, Mantell's plane crashed after he reported that he was at an altitude of 22,000 feet. Although investigators concluded that he had blacked out from lack of oxygen, speculation persisted that Mantell had been shot down by an extraterrestrial spacecraft. While it is now believed that Mantell was chasing a Skyhook balloon outfitted with a camera (later used for secret reconnaissance over Iron Curtain countries), the more colorful and exciting extraterrestrial rumors were hard to quash. This was only the beginning of many hard lessons concerning the UFO phenomenon. The Air Force would investigate this and a growing number of UFO reports through Project Sign, set up on January 22, 1948; Project Grudge, set up on December 16, 1948; and finally Project Blue Book, set up in March 1952, and continuing for 17 years.

Fig. 5.1. Cover from Hugo Gernsback's *Science Wonder Stories* for November 1929, depicting powerful aliens in a saucerlike spacecraft, foreshadowing the beginning of the modern UFO controversy two decades later. Copyright 1929 by Gernsback Publications, Inc.

The extraterrestrial hypothesis first emerged officially in Project Sign, where an "Estimate of the Situation" in late 1948 concluded that the UFOs were of extraterrestrial origin. But General Hoyt S. Vandenburg disagreed, and the report was returned, declassified, and burned. For the time being, the extraterrestrial hypothesis lost ground in the Air Force, but Project Sign's final report still left open the possibility that the UFO phenomenon might be something extraordinary and extraterrestrial. However, because it lacked the facts for an objective assessment, the study labeled the ideas of extraterrestrial space ships or atomic-powered aircraft "largely conjecture." In an appendix to the report on "the likelihood of a visit from other worlds as an engineering problem," James E. Lipp of the Rand Corporation placed thousand-to-one odds against the existence of higher life forms in our solar system and concluded that although space travelers from neighboring stars were much more likely than spaceships from Mars, this would require propulsion systems as yet unconceived on Earth. Visits from space were possible, he concluded, but they were "very improbable," and the actions attributed to flying saucers in 1947 and 1948 "seem[ed] inconsistent with the requirements for space travel."

When Project Grudge replaced Project Sign at the end of 1948, it had a less open-minded strategy. In the words of the historian David Jacobs, "New staff people replaced many of the old personnel who had leaned toward the extraterrestrial hypothesis. In the future, Sign personnel would assume that all UFO reports were misidentifications, hoaxes, or hallucinations." Project Grudge shifted the focus from explaining an unusual phenomenon in the atmosphere as something real to explaining it as illusion. A *Saturday Evening Post* article stated the new Air Force philosophy, backed up by Nobel Prize-winning chemist Irving Langmuir, a Project Sign consultant, whose advice to the Air Force on UFOs was to "Forget it!"

Project Grudge did, however, take one step that would be of profound importance to UFO history. It hired J. Allen Hynek (Fig. 5.2), an astronomy professor at nearby Ohio State University, to examine possible astronomical explanations for UFOs. Hynek (1910–1986) had come to Ohio State immediately after graduating with a Ph.D. from the University of Chicago in 1935. In 1956 he would go on to become the associate director of the Smithsonian Astrophysical Observatory, and 4 years later he became chairman of the Astronomy Department at Northwestern University. The year 1949 marked the beginning of a lifelong association with the UFO problem, culminating with his founding of the Center for UFO Studies in 1973.

In his capacity as author of many articles and books, and as a consultant to the Air Force on UFOs, Hynek would make a long voyage from skepticism to openness toward the extraterrestrial hypothesis. For now, in 1949, he was entirely skeptical, even though he admitted that 33 percent of all sightings could not be explained by astronomical or other physical phenomena. The Final

Fig. 5.2. J. Allen Hynek, one of the few scientists to play a central role in the UFO debate, began as a strong skeptic but gradually came to believe that UFOs required serious study. Courtesy J. Allen Hynek Center for UFO Studies.

Report of Project Grudge, issued in August 1949, attributed those sightings to psychological causes. With that conclusion Project Grudge was officially terminated, although some activity continued over the next 2 years.

It was thus not the Air Force that disseminated the extraterrestrial hypothesis, even if it raised the discussion of that hypothesis to an official level for the first time. Rather, this role was taken up by Donald Keyhoe (1897–1988), a retired Marine Corps major and pilot, who had managed the public tours of Admiral Byrd and Charles Lindbergh following their accomplishments and had written on Lindbergh, war, and aviation. Although Keyhoe was not a journalist, as a former chief of information at the Department of Commerce, he knew what media exposure meant and how to package an idea successfully for public consumption. Keyhoe became interested in the UFO question in 1949, and after receiving little cooperation from the military, he surmised that the flying saucers really were from outer space. This he made the thesis of a 1950 article, "The Flying Saucers Are Real," which, when expanded into several books, made Keyhoe the chief early exponent of the extraterrestrial hypothesis. But Keyhoe was peculiarly selective in his arguments for this hypothesis. Recent research on the abundance of planetary systems was juxtaposed with Lowell's idea of a Martian civilization, an idea rejected by most scientists for more than 40 years. Moreover, Keyhoe tended to explain anything mysterious by the extraterrestrial hypothesis. Radio signals of unknown origin might be attributable to navigation beacons by spacefaring aliens. Such explanations reveal Keyhoe's major weakness: throughout his book he had little idea of what constituted good scientific evidence or how to make sound scientific inferences from the evidence. Neither, unfortunately, did the public.

There things might have rested – a far-out and passing idea used to explain a few isolated instances – had not nature once again intervened to keep events rolling. In September 1951 an Air Force pilot reported a UFO over Ft. Monmouth, New Jersey, and the Air Force decided to revitalize its UFO investigation, naming Captain Edward J. Ruppelt to head the new study. But this was only the beginning. Within months the United States was enveloped in its first full-blown modern wave of flying saucers, and in March 1952 the Air Force responded by upgrading its investigation, now code named Project Blue Book. The Air Force at this time renamed flying saucers "unidentified flying objects," and Ruppelt stated that while the Air Force has never denied the possibility of interplanetary craft, "UFO reports offer absolutely no authentic evidence that such interplanetary spacecraft do exist." By year's end the number of reports topped out at just over 1500, with 536 reports in the month of July alone. In that month a series of sensational sightings over the Washington, D.C., area, including radar sightings from National Airport, evoked widespread interest among the public and concern in the government. Once again, while the Air Force investigated, the scientists dawdled, and the press took the lead in the search for explanations. This was to prove a fatal lapse on the part of the scientists, whose investigation at this point might have changed the subsequent sensationalist history.

As matters stood, though, the press was left to investigate, and an article by H. B. Darrach, Jr., and Robert Ginna in *Life* magazine asking "Have We Visitors from Space?" soon rivaled Keyhoe's 1950 article for popular influence. Among the propositions the article considered "firmly shaped by the evidence" were that "disks, cylinders and similar objects" have been present in the atmosphere of the Earth; that these objects "cannot be explained by present science as natural phenomena – but solely as artificial devices, created and operated by a high intelligence," and that no technology on Earth could account for the performance of these devices. In short, in the opinion of the authors, the extraterrestrial hypothesis was the only plausible explanation.

Clearly, leaving a subject requiring scientific analysis to reporters was a dangerous game. On the other hand, tackling such a controversial subject could also be dangerous to the career of any scientist, especially one not well established. Perhaps for these reasons – and because Hynek, as an Air Force consultant, seemed unwilling or unable to write for public consumption – one

Fig. 5.3. Donald Menzel, seen here about 1964 as Harvard College Observatory director. Menzel believed UFOs had prosaic explanations and remained a skeptic of the extraterrestrial hypothesis throughout his life. Harvard College Observatory photo.

prominent astronomer, and only one, did step forward in the early 1950s as a scientific spokesman about the UFO phenomenon. Donald Menzel (1901–1976), who became acting director of Harvard Observatory in 1952 and served as its director from 1954 to 1966, was one of the pioneering astrophysicists in the United States. Having received his doctorate from Princeton in 1924 under Henry Norris Russell, Menzel (Fig. 5.3) went on to a distinguished career in solar physics and more general problems of astrophysics at both Lick (1926–1932) and Harvard observatories (1933–1971). Menzel clearly enjoyed mystery, challenge, and imagination, and in his early days had published regularly (under assumed names) in science fiction magazines. Moreover, during World War II he had studied the effect of atmospheric irregularities on radar waves, and therefore felt himself particularly well suited for analyzing radar sightings of UFOs such as those occurring in 1952 over Washington, D.C. In many ways an ideal scientist to study UFOs, Menzel had a reputation that was above reproach, and association with the subject of UFOs was unlikely to harm his well-established career. As the reports of UFOs grew after 1947, Menzel recalled, he studied them with an eye toward natural explanations and early on concluded "with a slight feeling of disappointment" that the fly-ing saucers were not extraterrestrial but instead mundane objects. Menzel maintained this position unchanged throughout his life.

As early as April 1952 Menzel had presented his ideas on UFOs at a con-ference with Air Force officials in Washington. From then on, in articles, interviews, or books, he preferred radar ghosts or mirages, and sometimes as-tronomical objects, as explanations for the majority of UFO reports. As he stated in his first book on the subject, *Flying Saucers* (1953), Menzel proposed to combine the skill of the detective with the logic of the scien-tist in analyzing UFO reports. He would follow Sherlock Holmes's advice that "when you have eliminated the impossible, whatever remains, however improbable, must be the truth." While this alone might lead to the adop-tion of the hypothesis of alien spacecraft, the scientific side demanded that one always choose the simplest of a variety of hypotheses that will explain a given set of observational data. A scientist, Menzel went on, looks for explanations but does not arbitrarily invent forces that make explanation unnecessary. Extraterrestrial beings, in Menzel's view, were too complex a hypothesis, to be considered only if simpler, natural explanations failed. He thought natural explanations could account for the observed phenom-ena even though he believed in the possibility of extraterrestrial intelligence. Most other astronomers did not give their opinion in public but undoubtedly agreed with Menzel.

Even the Central Intelligence Agency, however, seems not to have been sat-isfied with the explanations of Menzel or the investigation of the Air Force. In December 1952, fearing that UFO reports could be used to clog commu-nications channels during times of national emergency, the agency secretly appointed a panel of five scientists to evaluate any threat that UFOs posed to national security. H. P. Robertson, Samuel A. Goudsmit, Luis Alvarez, Lloyd Berkner, and Thornton Page were all respected scientists who had worked on government projects during World War II. The panel met for 3 days in mid-January 1953 and reviewed some 75 case histories of sightings. They consid-ered the extraterrestrial hypothesis and noted that "none of the members of the Panel were loath to accept that this earth might be visited by extraterres-trial intelligent beings of some sort, some day. What they did not find was any evidence that related the objects sighted to space travelers." Their re-port, the Robertson Report, concluded that all sightings could be explained by prosaic causes and suggested that, in the interest of keeping communica-tions channels unclogged, the UFO phenomenon should be stripped of any aura of mystery and the public reassured that UFOs represented nothing in-imical. But because the report was classified secret, the conclusions of the few scientists who officially considered the UFO question remained unknown to the public until it was partly declassified in 1966.

Although the Robertson committee carried out its responsibility in determining that UFOs were not a threat to national security, it left open the question of what exactly they were. From the scientific point of view, Menzel's explanation that UFOs were mirages remained the sole attack of science on the subject. Even if not all scientists were convinced, they remained silent. But J. Allen Hynek, the one astronomer aside from Menzel who knew most about the data, had been growing increasingly skeptical of Menzel's explanations. Finally breaking his silence on the subject, in 1953 Hynek wrote that "in the absence of any universal hypotheses for the phenomena which stimulated these [UFO] reports, it becomes a matter of scientific obligation and responsibility to examine the reported phenomena seriously, despite their seemingly fanciful character." He was not proposing to accept the extraterrestrial hypothesis, but he was beginning to wonder if some kind of natural phenomenon was at work. Beyond this Hynek remained silent, understandably tied to his job in mainstream astronomy in academia.

In the absence of further serious scientific study, the claims of unqualified individuals received more attention than was warranted. George Adamski, a handyman at a hamburger stand near Mt. Palomar, claimed that he had made actual contact with alien spacecraft and their occupants and had even journeyed on their spacecraft. His book, *Flying Saucers Have Landed* (1953), as well as others that followed, were probably more widely known to the public than was Menzel's sole attempt to discredit the UFOs. They brought new scientific disrepute to the subject of UFOs, undoubtedly reinforcing scientists' belief that they should stay out of the controversy. Scientists seemed as unwilling to distinguish a potentially credible UFO phenomenon from Adamski's claims as the public was to separate scientific belief in extraterrestrials from UFOs.

With scientific study seemingly more remote than ever, the public took matters into its own hands. One result was the founding of private UFO organizations to collect and analyze reports – as if the public, fed up with scientists' inattention to stones falling from the sky, set up its own organization to study the phenomenon. With little or no scientific guidance, such organizations were subject to wild variations in the quality of their study, yet they filled a vacuum the scientists refused to occupy. In 1952 Coral and Leslie Lorenzen founded the Aerial Phenomena Research Organization (APRO) to study further the UFO phenomenon in the belief that the extraterrestrial hypothesis was plausible and that extraterrestrials might be engaged in systematic study of the Earth. And in 1956 the National Investigations Committee on Aerial Phenomena (NICAP) was founded, under the direction of Keyhoe until 1969.

At the same time, the Air Force continued its study, as detailed in Edward J. Ruppelt's *Report on Unidentified Flying Objects* (1956). The continuing Air Force investigation, and Ruppelt's own opinion that many unexplained UFO sightings remained, gave the subject respectability just at the dawn of the Space Age, even as science remained publicly aloof from the subject. Even the dawn of the new era, however, did not inspire an immediate scientific reaction to UFOs. But the opening of the Space Age in 1957 did coincide with a considerable UFO wave (Table 5.1), with the peaks of the wave, as reported to ATIC, occurring in October and November, just after the launch of Sputnik. Whether or not the Space Age precipitated this wave is an important historical question; it is surely significant but ambiguous that as more people briefly turned their attention to the skies, the UFO phenomenon also increased in number. Believers in a real UFO phenomenon could interpret this effect as more observers seeing what was always there if only one looked carefully enough, while skeptics could interpret the increase as public inexperience or gullibility about the heavens.

There were undoubtedly many reasons for the continuing public interest in UFOs. Writing in 1959, the psychologist Carl Jung argued in his *Flying Saucers: A Modern Myth of Things Seen in the Sky* that UFOs were a response to the lack of security and world peace. He also made a telling observation about media responsibility for UFOs. Based on his experience in 1958 of a press report that he was a "saucer-believer" and its refusal to correct the erroneous statement, Jung believed "one must draw the conclusion that news affirming the existence of Ufos is welcome, but that skepticism seems to be undesirable. To believe that Ufos are real suits the general opinion, whereas disbelief is to be discouraged. This creates the impression that there is a tendency all over the world to believe in saucers and to want them to be real, unconsciously helped along by a press that otherwise has no sympathy with the phenomenon." Other psychological interpretations of the UFO phenomenon were more closely related to the extraterrestrial hypothesis. As Ruppelt had perceptively said in his Air Force study in 1956, the "will to see" may have "deeper roots, almost religious implications, for some people. Consciously or unconsciously, they want UFO's to be real and to come from outer space."

Those believing that the extraterrestrial hypothesis was more than a psychological yearning, however, were finally supported by the French astronomer and computer scientist Jacques F. Vallee (1939–). Vallee received his B.S. in mathematics from the Sorbonne in 1959, his M.S. in astrophysics from Lille University in 1961, and his Ph.D. in computer science from Northwestern University in 1967, where he came into contact with Hynek. Impressed by the 1954 wave of UFO sightings in France and influenced by Aime Michel, the leading UFO researcher in that country, Vallee took up the subject of UFOs and in 1965 offered his own appraisal of the subject in *Anatomy of a Phenomenon*. Vallee challenged the scientific method of Menzel and others, finding the prevailing attitude too limited, the data sample too small, and the techniques too narrow. Vallee believed one should not invoke a new natural phenomenon to explain UFOs if the extraterrestrial hypothesis was plausible; extraterrestrials seemed to be his working hypothesis for the UFO phenomenon, and much of his book therefore aimed to show the plausibility of advanced civilizations, interstellar travel, and UFOs as alien spaceships. Vallee would persist in his UFO career, joining up with Hynek and becoming one of the leaders in the field. The views of both would eventually evolve beyond a nuts-and-bolts alien spaceship interpretation of UFOs in a synergistic effort that symbolized the promise and the perils of a scientific approach to UFOs. But Vallee's book was an early indication that even when scientists took up the UFO problem, not only a phenomenon was at issue, but also the scientific method for analyzing it.

By contrast, Menzel had not changed his mind in the slightest and was not about to change his method. In *The World of Flying Saucers: A Scientific Examination of a Major Myth of the Space Age* (1963), he still argued that all UFOs were explainable in natural terms, despite professing an open mind on the question of extraterrestrial life. Although one must remember that the discovery of meteorites was greeted with disbelief, he cautioned, "an open mind does not mean credulity or a suspension of the logical faculties that are man's most valuable asset." Although visits to and from other worlds may occur in the future, Menzel concluded, no evidence suggests that such visits have yet begun. Nevertheless, over the next few years, that hypothesis would finally receive more serious scientific scrutiny.

5.2 THE PEAK OF THE EXTRATERRESTRIAL HYPOTHESIS, 1965–1969

During the second half of the 1960s, the extraterrestrial hypothesis of UFOs was the subject of more attention from mainstream science than at any time before or since. Whereas for 17 years since 1947 the Air Force had been the primary investigator of the UFO phenomenon and the media the primary purveyors of the extraterrestrial hypothesis, from 1965 to 1969 at least a few in the U.S. scientific community finally took the subject seriously, in part because of pressure from the U.S. Congress. Whereas before 1965 one could not speak of a spectrum of scientific opinion on UFOs, because only a few scientists such as Hynek and Menzel had ventured an independent opinion, by 1970 that spectrum was well populated. If the media and the Air Force dominated the period to 1965, in the second half of the 1960s it was the turn of the scientists.

Amid the political and intellectual turmoil that characterized the 1960s, the broader scientific and technical landscape had also changed. With the first radio search for extraterrestrial intelligent signals, the announcement of a possible planet around Barnard's star, and Shklovskii and Sagan's popular book *Intelligent Life in the Universe* (1966), extraterrestrials became the subject of renewed interest. Moreover, the United States was halfway to President Kennedy's goal of landing a man on the Moon. Space was becoming a real place in the popular mind, and it made sense that if Earthlings were traveling in space, aliens might too. All of this background was combined, often through loose thinking by association, in such popular books as Frank Edwards's *Flying Saucers – Serious Business* (1966), which sold in excess of a million copies and was still in print in the last decade of the century. With its large section on the possibility of life in outer space, Edwards's book and others became best-sellers in part because they favored the extraterrestrial explanation for UFOs, a sure bet to catch the public imagination. For those who thought more critically and appreciated the immense distances likely to exist between civilizations, the link between extraterrestrial intelligence and UFOs was not so obvious, though the immensity of time argued that advanced civilizations had billions of years to colonize the universe and therefore might well be in Earth's vicinity. Argument without critical observation could go on forever, but positive evidence was the crucial factor for any explanation of the UFO phenomenon. And that is where the scientists could potentially be of most help.

The precipitating factor for renewed interest in the extraterrestrial hypothesis in the 1960s was a new wave of UFOs, which filled the skies from July 1965 to mid-1967. Instead of the usual 30 or so reports per month, hundreds were reported; the ATIC received 887 in 1965 alone. In part because of the changed scientific climate, this time the reaction was different from that in 1952. No longer was it possible for a 1-week panel like the Robertson committee to undertake a limited study and pronounce that all was well. Instead, the trajectory of events led from the Air Force to congressional hearings; a 2-year university study headed by the physicist Edward U. Condon, which precipitated a mixture of scientific reactions; and a symposium on the subject sponsored by the prestigious American Association for the Advancement of Science (AAAS). The trajectory began with increased public interest in UFOs and in the Air Force's previous investigations, prompting Hynek to propose to the Air Force a panel of civilian scientists to investigate UFOs. As a result, in the fall of 1965, the Air Force set up the Ad Hoc Committee to Review Project Blue Book, composed mostly of members of the Air Force Scientific Advisory Board. Chaired by Brian O'Brien, the group met on February 3, 1966. It found that of 10,000 sightings reported and classified over the previous 19 years, "there appears to be no verified and fully satisfactory evidence of any case that is clearly outside the framework of presently known science and technology." Hedging its bet, the committee nevertheless recommended a strengthening of the Air Force program by contracts to universities to study the problem.

The O'Brien committee's recommendation would soon be carried out, but not before other events intervened. On March 20 and 21, 1966, more than 100 witnesses reported UFOs in the cities of Dexter and Hillsdale, Michigan. The Air Force sent Hynek to investigate; his tentative explanation was that swamp gas had spontaneously ignited, causing a faint glow. This explanation was believed by virtually no one; Hynek was ridiculed, and the press had a field day over the issue of the inadequacy of scientific explanations of UFOs. Hynek later recalled that this was a watershed experience for him: while he had had private misgivings as early as 1953 about Blue Book policies, "it wasn't until after the 'swamp gas' incident that I said, 'I've had it! This is the last time I'm going to try to pull a chestnut out of the fire for the Air Force.'"

Hynek's transformation was not the only result of the Michigan sightings; his explanation was considered so inadequate that Michigan Congressmen Weston Vivian and Gerald Ford called for congressional hearings. On April 5, 1966, the House Armed Services Committee held open hearings on the subject of UFOs. Air Force Secretary Harold Brown, Blue Book head Hector Quintanilla, Jr., and Hynek were invited to testify. Secretary Brown brought with him statistics showing that of the 10,147 UFO sightings reported to the Air Force between 1947 and 1965, the Air Force had identified 9,501 as natural phenomena wrongly interpreted. The remaining 646, Brown asserted – carefully choosing his words – were "those in which the information available does not provide an adequate basis for analysis, or for which the information suggests an hypothesis but the object or phenomenon explaining it cannot be proven to have been here or taken place at that time." In short, Brown made it clear that neither he nor anyone involved with Project Blue Book believed UFOs to be extraterrestrial. But Hynek, invoking cases in the history of science such as fossils, X-rays, and meteorites, called for a civilian panel of physical and social scientists to examine whether UFOs were a significant scientific problem. Out of the hearings came a decision to implement the O'Brien committee's recommendation for further scientific investigation. In April the Air Force moved to implement such an investigation, and in October, after a difficult search, it named (on Menzel's recommendation) Edward U. Condon (Fig. 5.4) as its head.

Edward Condon (1902–1974) was a physicist who had made significant contributions to quantum mechanics and had experience in academia, industry, and government. At the urging of the secretary of commerce, in 1945 President Harry S. Truman appointed Condon director of the National Bureau of Standards. When Condon became one of the accused during the infamous House Un-American Activities Committee hearings, in 1951 he left the bureau for Corning Laboratories. In about 1955 he became chairman of the Physics Department at Washington University at St. Louis, and later came to

Fig. 5.4. Edward U. Condon, seen here in his office at the University of Colorado in the early 1970s. A physicist with a long and distinguished career, he headed a U.S. Air Force study that became a landmark in UFO history. Courtesy National Institute for Standards and Technology.

Boulder, Colorado, as a professor and fellow of the Joint Institute for Laboratory Astrophysics. It was here, at the University of Colorado, that the study would be carried out.

It was in this context, during an "incautious moment" as one of his biographers has said, that Condon accepted the task of heading the Air Force investigation of UFOs. By his own account Condon was not eager: "I did not want to do the UFO study," he wrote several years later to Carl Sagan, "but was talked into it in August 1966 by staff of the U.S. Air Force Office of Scientific Research, largely on the basis of appeals to duty to do a needed public service. . . ." By another account, during a losing campaign to become a regent of the university, Condon had stressed the need for more federal research money, and the $300,000 the Air Force was offering (eventually supplemented with another $200,000) was not to be taken lightly. Undoubtedly both the scientific challenge of the problem and the federal money were factors in Condon's acceptance of the task.

Even before the Condon study began, the new wave of UFOs brought some scientists out of the closet, and Condon knew he would have to deal with them. Particularly important was the work of James McDonald, an atmospheric physicist at the Department of Meteorology and the Institute of Atmospheric Physics at the University of Arizona in Tucson. A member of the National Academy of Sciences, McDonald had been casually interested in UFOs for many years. A particular sighting in Tucson in March 1966, as well as the Michigan sightings during the same month, led him to study the subject much closer; by mid-1966 he had visited Hynek and Blue Book officials and had come to believe that the extraterrestrial hypothesis was the best explanation for UFOs. In 1967 he presented a detailed lecture in Washington, D.C., in which he argued that UFOs were "the greatest scientific problem of our times," possibly explainable only by the extraterrestrial hypothesis. As the Condon committee was undertaking its work, McDonald's ideas were publicized in the national press. In private, McDonald was sparring with his colleague at the University of Arizona, the astronomer Gerard P. Kuiper. While Kuiper agreed that the UFO phenomenon had been given "superficial treatment," he argued that extraordinary evidence would be required to prove the extraterrestrial hypothesis, and he stated that UFOs were probably unexplained terrestrial atmospheric phenomena. McDonald remained unconvinced.

The new wave of UFOs also brought great interest in the popular press. Science editor John Lear's series of articles in the fall of 1966 in the *Saturday Review* focused attention on the phenomenon. Lear not only brought to readers' attention the ideas of Jacques Vallee on UFOs and those of Carl Sagan on extraterrestrial intelligence, he also managed to obtain an edited version of the Robertson committee's report (not officially released until several years later) and apprised the public of the Air Force–sponsored study. Lear's articles also brought to public attention a more sensational and disturbing aspect of the UFO debate during this time – the story of sightings in New Hampshire and the claimed abduction of Betty and Barney Hill by occupants of a spacecraft that had allegedly landed in that state, a story documented and popularized by Lear's fellow editor at the *Saturday Review,* John Fuller.

With such activity raging around it, all looked to the Condon study, which became the focus of UFO debate in the second half of the 1960s. The study is of interest both for how it was undertaken and for the conclusions reached. On the first point, there is no doubt that the execution of the study was somewhat tumultuous. Disagreement might have been expected, based on the widely varying opinions as to the scientific merit of the subject, but no one could have seen the tumult that developed. Condon's original proposal seemed straightforward enough. He would spend about half of his time as director of the UFO project. A University of Colorado administrator, Robert Low, would be the scientific coordinator and principal investigator. They would be joined by several colleagues in the university's Psychology Department, including its chairman, Stuart Cook, and David R. Saunders. The study would also have the benefit of members of two nearby national institutions: the expertise of the astronomer Franklin Roach of the Environmental Science Services Administration (ESSA) and the cooperation of the National Center for Atmospheric Research, headed by the astronomer Walter Orr Roberts. The project, which would eventually cost about $500,000 over 2 years, would examine the files of the Blue Book panel, talk to its staff members, and conduct interviews, field studies, and tests. The contract (executed between the University of Colorado and the Air Force Office of Scientific Research) indicates that Condon realistically expected no single explanation to emerge. It would be surprising if an unambiguous physical interpretation would be found for all cases, he noted, and more reasonable to expect "that the phenomena reflect a variety of perceptual and cognitive processes superimposed on a variety of physical stimuli." The methodology would "lean in the direction of quantification and experimentation and away from the effort simply to enlarge the already substantial body of opinion and impression," although Condon explicitly recognized that the physical scientists might differ in their methodologies from the behavioral scientists.

It all sounded very promising, but during the course of 1967–1968, the study was wracked by internal dissension and outside controversy. The details have been examined elsewhere, but the controversy was fueled at least in part by the question of how to approach the extraterrestrial hypothesis. One faction, led by the psychologist Saunders, wanted to concentrate on that hypothesis; the other, led by Low, believed the solution lay in the psychology of the witnesses. When an internal memorandum written by Low was leaked to McDonald by Saunders and was interpreted as biased against a fair study, Condon fired Saunders and another staff member. By the end of the study, only 3 of the 12 original staff members remained. When in the summer of 1968 these problems hit the pages of *Science,* the journal of the AAAS (an organization that Condon had headed), Condon refused to talk to its reporters and resigned his AAAS membership.

A few days later, the House Committee on Science and Astronautics, aware of the controversy over the continuing Condon study, held hearings on the UFO problem, broadening still more the scientific participation. Indiana Representative J. Edward Roush, who had shown a personal interest in the UFO problem and even corresponded with Kuiper about it, chaired the symposium. The testimony that followed from Hynek and McDonald, as well as from the Cornell astronomer Carl Sagan and others, surely gave Roush and his congressional colleagues a lesson on how inexact science could be. Hynek argued that one should not confuse "psychically unbalanced individuals and pseudoreligious cultist groups" who hoped for salvation from UFOs with those who sought scientific explanations. McDonald, criticizing the prosaic explanations of Menzel and others, took seriously the possibility "that we are dealing with surveillance from some advanced technology." And Carl Sagan, having just 2 years earlier published with Shklovskii *Intelligent Life in the Universe,* found the extraterrestrial hypothesis for UFOs not at all persuasive. He was much more impressed with the psychological explanations of Jung and others, pointing to "unfulfilled religious needs." Sagan therefore supported a moderate investigation of UFOs but suggested that it would be much better for Congress to support the spacecraft investigations of life and radio searches for intelligent signals.

In the midst of all the internal dissension and external scrutiny, the Condon study ground on. The *Final Report of the Scientific Study of Unidentified Flying Objects* was a massive multiauthored volume, delivered to the Air Force in November 1968 and released in January 1969. The bulk of its 967 pages in the paperback edition consisted of a description of the work of the project, case studies, and scientific explanations. Condon's conclusions, which opened the report, disappointed many a UFO enthusiast. Despite the committee's best efforts, Condon stated, "Our general conclusion is that nothing has come from the study of UFOs in the past 21 years that has added to scientific knowledge. Careful consideration of the record as it is available to us leads us to conclude that further extensive study of UFOs probably cannot be justified in the expectation that science will be advanced thereby." Although this was enough for officials who wished to put an end to the subject, Condon's further elaboration is more revealing. He went on to talk about how science works and how each scientist must make a personal decision as to what research will be most fruitful. Scientists should not uncritically accept the conclusion of the report, he stated, urging that if new ideas for "clearly defined, specific studies" were forthcoming, they should be supported. Condon made it clear that he did not believe that any extensive further effort should be made to study UFOs at the present time, but that "this may not be true for all time."

Condon's "summary of the study" addressed the extraterrestrial hypothesis. Distinguishing between the extraterrestrial hypothesis (ETH) that UFOs may represent alien spacecraft and the extraterrestrial actuality (ETA) that such spacecraft are an observational fact, Condon and his colleagues found that "no direct evidence whatever of a convincing nature now exists for the claim that any UFOs represent spacecraft visiting Earth from another civilization." Decisive observations are sometimes hard to come by, Condon emphasized, and therefore progress in science can be painstakingly slow. He left no doubt about the kind of hard evidence he would like to see: "The question of ETA would be settled in a few minutes if a flying saucer were to land on the lawn of a hotel where a convention of the American Physical Society was in progress, and its occupants were to emerge and present a special paper to the assembled physicists, revealing where they came from, and the technology of how their craft operates. Searching questions from the audience would follow."

Most of the scientific world shared Condon's view of the importance of strong evidence. Therefore, when in October 1968 the Air Force requested the National Academy of Sciences to "provide an independent assessment of the scope, the methodology, and the findings of the study," the outcome was not surprising. The president of the National Academy (Frederick Seitz, a student of Condon at Princeton in 1930) appointed a panel chaired by the Yale astronomer and former U.S. Naval Observatory Scientific Director Gerald M. Clemence. After about 6 weeks of study, the panel agreed with all the findings of the Condon committee, concluding that "on the basis of present knowledge the least likely explanation of UFOs is the hypothesis of extraterrestrial visitations by intelligent beings," and further agreed that "no high priority in UFO investigations is warranted by data of the past two decades."

The reaction of the broader scientific community, however, would not be so unanimous. To be sure, there were those who approved of the Condon Report. *Science,* the official journal of the AAAS, emphasized the report's "massive documentation," its finding that seemingly hard evidence was often faulty, and the fact that most cases tended to be explicable when enough evidence was available. The MIT physicist Philip Morrison – who in 1959 had proposed the radio search for extraterrestrial intelligence and was not known for his lack of imagination – wrote, "One comes away edified, amused, admiring and well satisfied. . . . Science is the stronger for this sincere and expert effort to deal with a public concern." And from Arizona, Gerard Kuiper congratulated Condon on an outstanding report.

But the dismissal of such a high-profile and publicly scrutinized phenomenon as UFOs was not going to be that easy. Scientific critics were vociferous from the time the Condon Report was released. David Saunders, the psychologist associated with the Condon project until his dismissal, released his book *UFOs? Yes! Where the Condon Committee Went Wrong,* an insider's story with disturbing revelations about the management of the project. James McDonald returned with renewed vigor to excoriate his colleagues for bad science, calling the Condon Report "$500,000 worth of the bum scientific advice the Air Force has been getting for 20 years." McDonald was disturbed at the mismatch between the results of the case studies and Condon's conclusion about the extraterrestrial hypothesis; he was the first of many to level this criticism. Condon seemed to have taken his sole aim to be the proof or disproof of the extraterrestrial hypothesis, McDonald complained, even though that hypothesis was only one among many.

McDonald was far from alone in his criticism. In particular, Condon must have been surprised and chagrined when an appointed UFO Subcommittee of the American Institute of Aeronautics and Astronautics (AIAA), a professional organization of 25,000 aerospace scientists and engineers no less distinguished than their AAAS colleagues, turned its back on Condon's conclusions after its own study. The subcommittee found it "difficult to ignore the small residue of well-documented but unexplainable cases which form the hard core of the UFO controversy." It noted that while the signal-to-noise ratio was very low, it could not be ignored: "The issue seems to boil down to the question: Are we justified to extrapolate from 0.99 to 1.00, implying that if 99% of all observations can be explained, the remaining 1% could also be explained; or do we face a severe problem of signal-to-noise ratio . . . ?" The subcommittee found that the Condon study had made no serious analysis in this direction, and therefore "strongly feels that, from a scientific and engineering standpoint, it is unacceptable to simply ignore substantial numbers of unexplained observations and to close the book about them on the basis of premature conclusions." To some this seemed a more objective review of Condon's study than that of the National Academy, and it was followed up with what the committee considered some of the best evidence. In a series of articles for its journal, *Astronautics and Aeronautics,* James McDonald and Gordon Thayer presented the details of two case studies that they concluded were still completely unresolved. This was to be McDonald's last article; on June 13, 1971, he was found dead from suicide in the desert near Tucson.

In the wake of the congressional hearings, the Condon Report, and the diverse reaction it precipitated, the capstone to scientific interest in UFOs during the late 1960s was a symposium on UFOs held in Boston in December 1969 in conjunction with the annual AAAS meeting. The very fact that the meeting was held at all was a testimony to changed times, but organizing it was not easy. In their introduction to *UFOs: A Scientific Debate* (1973), Sagan and Thornton Page noted that the meeting had been postponed in part because of opposition "from some very distinguished scientists" and was finally held only because of the steadfast courage of the AAAS president. While some scientists viewed UFOs as an unscientific subject analogous to Velikovsky and astrology, Sagan and Page argued that organizations such as the AAAS have the obligation to arrange for discussions on just these subjects that catch the public eye as a means of demonstrating the scientific method.

Among the 15 invited papers the now familiar positions of Menzel, McDonald, and Hynek were represented, as were the ideas of two of the Condon researchers, Franklin Roach and William K. Hartmann. Carl Sagan argued that the extraterrestrial hypothesis was possible but unlikely, while Frank Drake concluded that "Some aspects of perception work very well, some do well given certain qualifying conditions, and some fail completely." Philip Morrison, with his penchant for zeroing in on the significant issues, argued that the debate came down to the nature of scientific evidence. "Reproducibility" was not enough, for one could not reproduce an aurora or eclipse, nor was "hard evidence" enough. The prime requirement for responsible evidence, he held, drawing a parallel with the 19th-century acceptance of meteorites as extraterrestrial, was "independent and multiple chains of evidence, each capable of satisfying a link-by-link test of meaning." Neither the extraterrestrial hypothesis nor any other explanation of UFOs had multiple chains of evidence or a link-by-link test.

By the end of 1969 the spectrum of scientific opinion on UFOs was well filled. At one end McDonald, a few other scientists, and much of the media and public stood for the extraterrestrial hypothesis. At the other extreme Donald Menzel, Philip J. Klass, and the official Condon study believed science would not be further advanced by any study of UFOs – no matter what the hypothesis. In the middle Hynek and Sagan, the AIAA committee, and a mostly silent number of astronomers felt that further study might be beneficial, if difficult. As for the Air Force, it ended Project Blue Book on December 17, 1969. Less than a year after the end of his landmark study, Condon gave a brief and light-hearted (some would say flippant) account of his experiences during the study, emphasizing some of the stranger stories he had encountered. It was his last published word on UFOs before his death 4 years later. Though he had charted a distinguished career in quantum physics, the public – and not a few scientists – would remember him primarily for the lasting stamp he put on the UFO debate during 2 years at the height of the controversy in which extraterrestrials had played a dominant – perhaps too dominant – role.

5.3 AFTERMATH: THE NATURE OF EVIDENCE AND THE DECLINE OF THE EXTRATERRESTRIAL HYPOTHESIS IN PHYSICAL SCIENCE

By 1970 the Condon study was over, the U.S. Air Force was out of the UFO business, and the U.S. Congress was apparently pacified, or at least preoccupied with more immediate concerns such as Vietnam, whose reality – in contrast to that of UFOs – was unquestioned. In retrospect, a quarter-century after the beginning of the modern UFO era with the Kenneth Arnold sightings in 1947, we see in the early 1970s the beginnings of a distinct transformation of the UFO debate. UFO sightings were not over – the 1973 national wave kept the controversy going, as did regional waves thereafter – but the hypothesis that UFOs represented alien spacecraft was entering a period of decline, at least from the point of view of physical science and official government attention. Whereas in the 1960s establishment science finally addressed the subject of UFOs with an official study, a National Academy of Sciences review, an AAAS symposium, the persistent prodding of one member of the National

Academy (McDonald), and occasional reports and reviews in its mainstream journals like *Science,* such activity rapidly decreased in the 1970s. One may argue whether or not the scientific community had taken the extraterrestrial hypothesis seriously enough in these activities of the 1960s, or whether it should have broadened the scope of its scientific method in addressing the problem, but the historical fact remains that in the aftermath of the Condon study the hypothesis declined in mainstream science, as measured by scientific attention and discussion in established scientific journals. At the same time, this was decidedly not the case outside science, where the extraterrestrial hypothesis of UFOs remained as alive and well as ever. In this section we examine the reason for the decline of the extraterrestrial hypothesis in science and its continuing viability in American society notwithstanding the view of science.

At least three circumstances joined in precipitating the decline of the extraterrestrial hypothesis of UFOs in mainstream science. First, even taking into account the failings of the Condon study, no incontrovertible evidence was produced in favor of the hypothesis, and its champions realized that to maintain the extraterrestrial theory lacking such evidence was an obstacle to further study of the UFO phenomenon. Second, "new wave" theories of UFOs led away from the extraterrestrial hypothesis and toward ever more ethereal ideas associated with the New Age movement that grew in the last quarter of the 20th century, confirming to scientists that this was a subject to be avoided. Finally, the rise of claims that, if true, would have proved the extraterrestrial hypothesis – ancient astronauts, spaceship crashes, contactees, and abductees – were based on evidence (such as statements under hypnosis) that most scientists could not accept, thereby bringing the entire extraterrestrial hypothesis into disrepute. Pervading all of these factors was the question of scientific method and the nature of evidence, questions that mean a great deal to working scientists but often all too little to the public. Therein is the entering wedge for the dichotomy of opinion between scientists and the public that once again grew wider during the last quarter of the century.

J. Allen Hynek is a prime example of one who was open to the extraterrestrial hypothesis but realized he lacked the proof in any standard sense of scientific method. Hynek's first book-length treatment of the problem, which quickly became a classic in the field, clearly demonstrates why he moved away from the alien as an explanation for the phenomenon he had grappled with for almost a quarter-century. *The UFO Experience: A Scientific Inquiry* (1972) was perhaps a book Hynek should have written long before, considering his lengthy association with the Air Force study. In it he introduced his "strangeness-probability" diagrams as a measure of the importance of particular UFO cases. He classified UFO reports into six prototypes, three of which were in the "distant" category (nocturnal lights, daylight discs, and radar and visual reports) and the other three of which were "close encounters." Hynek showed how far he had come since his skeptical days when he analyzed in a serious way not only the distant category but also the close encounters, bringing into the UFO lexicon the phrases "close encounters of the first kind" (a close experience without physical effects), "close encounters of the second kind" (measurable physical effects), and "close encounters of the third kind" (animated creatures reported). In the chapter titled "Science Is Not Always What Scientists Do," Hynek noted (as had McDonald) that the Condon Report contained many unexplained cases, which Condon chose to ignore in his summary and recommendations; that by choosing to equate UFOs with the extraterrestrial hypothesis, Condon wrongly rejected the entire UFO phenomenon; and that Condon himself was preoccupied with the weirdest cases, to the detriment of the better ones.

Significantly, Hynek pointed out that the extraterrestrial hypothesis could not be proved untrue, and deduced from this not that it was true (as some would), but that the Condon team had adopted the wrong methodology. While in his view the team should have tested the hypothesis "There exists a phenomenon, described by the content of the UFO reports, which presently is not physically explainable," they instead tackled a "hopeless task" – trying to test the extraterrestrial hypothesis. No crucial observation or experiment could prove the latter false, Hynek argued, while the former could have been proved false by explaining the cases that no one else had been able to explain. It was as if the field had been painstakingly narrowed to the good data, the signal separated from the noise, and then the signal thrown away. From his own experience, Hynek concluded that the UFO reports described a phenomenon "worthy of systematic, rigorous study," and (like the AIAA) he suggested that this be undertaken without dragging the speculative and emotional extraterrestrial hypothesis into consideration. Following up on this conviction, in 1973 Hynek founded the Center for UFO Studies as a repository for data and study of the UFO phenomenon.

The idea that the UFO problem might have a better chance of scientific study if stripped of its extraterrestrial connotations had some success. Hynek's plea for further study impressed at least one mainstream scientist, whose review of the book for *Science* stood in sharp contrast to that journal's endorsement of the Condon Report a few years before. Bruce Murray, Professor of Planetary Science at Caltech and a future director of the Jet Propulsion Laboratory, wrote, "On balance, Hynek's defense of UFOs as a valid, if speculative, scientific topic is more credible than Condon's attempt to mock them out of existence." And Murray did not stop there. He viewed Hynek's inability to obtain funds from NASA or the National Science Foundation to study UFOs as "a dismal symptom of the authoritarian structure of establishment science." Moreover, Murray castigated *Science* for treating the Condon Report as a news item rather than publishing a rebuttal. "From this juror's point of view at least," Murray concluded, "Hynek has won a reprieve for UFOs with his many pages of provocative unexplained reports and his articulate challenge to his colleagues to tolerate the study of something they cannot understand." He had not, however, won a reprieve for the extraterrestrial hypothesis, and we see the emphasis moving away from that theory even by those who advocated further study of the UFO phenomenon.

The same avoidance of the extraterrestrial hypothesis while pressing for further study is seen in the critique of the Condon Report by the astrophysicist Peter Sturrock of Stanford's Institute for Plasma Research. Sturrock pointed out that, despite the efforts of a few individuals, the great bulk of the scientific establishment still ignored the UFO problem, regarding it as unproductive and unrespectable. The reason for this negative attitude, he believed, was that scientists normally look for their hard data in scientific journals, which almost always refused to publish UFO articles based on the advice of peer reviewers. Furthermore, he emphasized that Condon's methodology seemed to demand a single convincing case, the normal method pursued by physicists, while the UFO phenomenon might demand a method more often adopted by astronomers: the laborious accumulation of data. No single observation in astronomy, he pointed out, proved Kepler's laws, or the shape of the galaxy, or stellar evolution. For this reason, Condon had rejected some of the data discussed by his own staff. Thus, Sturrock concluded, "the weaknesses of the Condon Report are an understandable but regrettable consequence of a misapprehension concerning the nature and subtlety of the phenomenon. . . . The substance of the Condon Report represents a persuasive case for the view that there is some phenomenological fire hidden behind the smoke of UFO reports, and the Report therefore *supports* the proposition that further scientific study of UFOs is in order."

Others, notably Allan Hendry at Hynek's Center for UFO Studies in the late 1970s, made similar pleas for drastically new methodologies to study the UFO phenomenon. By that time, several related themes were becoming clear among critics of Condon and previous studies: too much attention had been given to the extraterrestrial hypothesis; the UFO phenomenon deserved further study with a much broader focus; and any further study needed carefully to consider broadening its scientific methodology. The highly emotional issue of the extraterrestrial hypothesis came to be seen as an obfuscation by those supporting further study of the UFO phenomenon. Short of a spaceship landing in plain view, the hypothesis did not seem to be susceptible to normal scientific methods. Three decades of sporadic study had resulted in no incontrovertible proof. If, as Hynek had proclaimed, the extraterrestrial hypothesis was unfalsifiable, then by some definitions it was not a scientific hypothesis at all and attention should turn to other theories that were scientific.

UFO debunkers, however, were not about to let their prey slip away that easily. They used adherence to standard methodology not only to deny the extraterrestrial hypothesis, but also to deny that the UFO phenomenon was at all novel, significant, or outside the realm of normal explanation. To the end, Menzel held to his original credo that a scientist must always choose the simplest hypothesis, and using that method in his final book on UFOs published just before his death, he still believed all sightings could be explained on the basis of known natural phenomena. Robert Schaeffer, another debunker, found only one hypothesis left: "It is the familiar null hypothesis, the cornerstone of statistical analysis; UFOs, as a phenomenon distinct from all others, simply do not exist." Those who continue to believe otherwise, he concluded, are giving allegiance to a worldview other than science. Hynek, Murray, Sturrock, and others – all scientists themselves – would hardly agree; rather than discarding a phenomenon that did not fit within the normal canons of scientific evidence, they championed a broader view of science that could encompass the study of an unknown phenomenon.

Like the debate over the canals of Mars, the UFO controversy thus reveals the existence of many cultures of science, each with its own characteristics, approaches, and canons of scientific evidence (Table 5.2). Even those with broader views of the scientific method, however, agreed that while the UFO phenomenon should be pursued, there was no incontrovertible evidence for the extraterrestrial hypothesis and seemingly no way to obtain such evidence. In short, even as NASA laid plans to search for extraterrestrial intelligence by means of radio telescopes, the idea that such intelligence had traveled to Earth was not considered a fruitful scientific hypothesis.

The desire to broaden the scope and method of science was laudable but extremely difficult. Faced with finding alternatives to the extraterrestrial hypothesis, many unfortunately replaced it with even wilder theories. A second factor in the decline of the extraterrestrial hypothesis in mainstream science is the appearance of these "new wave" UFO theories – the notion that UFOs are metaphysical, supernatural, or interdimensional. While these theories were sometimes combined with the idea of extraterrestrial spacecraft, they began to supersede the nuts-and-bolts extraterrestrial hypothesis of spacecraft that traveled by the conventional laws of physics. Referring to the standard spaceship concept, Schaeffer found that "while the pace of UFO sightings quickened in the mid-to-late 1960s, the tenability of the extraterrestrial hypothesis was gradually deteriorating." By the mid-1970s, even former extraterrestrial hypothesis advocates such as Hynek and Vallee began to cast aside the standard spaceship hypothesis for new ones that involved parallel universes, interdimensional travel, and "astral planes." In "The Case Against Spacecraft" in his book *Messengers of Deception* (1980), Vallee argued that the spacecraft hypothesis no longer explained the facts of the UFO phenomenon. In a series

Table 5.2. *Spectrum of scientific cultures on the UFO question*

Explanation	Advocate	Characteristics
Prosaic natural phenomena (astronomical, balloons, etc.) or fraud	Menzel (1952–1976) Hynek (1950s–1965) Condon (1968) NAS panel (1969) Klass (1966–1990s) *Skeptical Inquirer*	Skepticism Caution "Simplest explanation" Need for strong evidence for theory
Potentially significant new phenomenon: further study needed	AIAA Committee (1968) Hynek (1966–1986) Murray (1972) Sturrock (1974) *Journal of Scientific Exploration* Vallee (1965 on)	Consider accumulation of data (no single confirmation) Extract signal from noise
Extraterrestrial hypothesis	McDonald (1966–1971) Hynek (1968 and sporadically) Vallee (1965–1980) *Journal of UFO Studies* (on occasion)	Risk takers Willing to speculate Willing to consider alternative methods, especially eyewitness accounts

of books in the late 1980s and early 1990s Vallee refined his thesis that "The genuine UFO phenomenon . . . is associated with a form of nonhuman consciousness that manipulates space and time in ways we do not understand." This form of consciousness "does not have to be extraterrestrial. It could come from any place and any time, even from our own environment. . . . The entities could be multidimensional beyond space-time itself. They could even be fractal beings. The earth could be their home port." Hynek also seemed sympathetic to such ideas, and if these once relatively conservative Ufologists were now attracted by such ideas, one can well imagine the hypotheses of those with lesser scientific backgrounds.

Although Vallee's ideas were by no means widely accepted, and although the extraterrestrial hypothesis continued to flourish at levels outside mainstream science, the very existence of such new wave ideas carried UFOs further from the realm of accepted science and into what most scientists considered pseudoscience. The scientific reaction to the broad realm of pseudoscience was inevitable, and as UFOs became more and more associated with that realm, so was the scientific reaction to UFOs. The Harvard astronomer Fred Whipple saw Menzel's last book, for example, as "a solid steppingstone out of the morass of unconscious yearning for supernatural intervention." That an extraterrestrial presence on Earth had become a part of the New Age philosophy was recognized even by Vallee: "To the New Age idealists, the announcement that aliens are here would bring the culmination of many decades of dreams. It would validate all their group meditations on mountaintops, the loving hopes, the prayers for peace. It would give all of us something to worship at a time when the leaders of our traditional religions have made fools of themselves, at a time when the younger generation has very few heroes it can look up to." The fact that aliens were devoutly desired by the New Age movement did not advance the extraterrestrial hypothesis one bit in the scientific community, nor were scientists impressed with Hynek and Vallee's abandonment of that hypothesis for ones even further removed from empirical confirmation. And if Vallee decried the New Age desire for alien salvation, his own ideas of UFOs surely found sympathy in the New Age philosophy.

Finally, even more weighty than the defection of Hynek, Vallee, and others to New Age explanations of the old UFO phenomena, the spectacular claims of new phenomena ironically contributed mightily to the downfall of UFO respectability among scientists. In particular, the rise of a belief in ancient astronauts, crashed spaceships, contactees, and abductees made the subject of extraterrestrial hypothesis disreputable; Vallee himself showed the absurdities of many of these claims in his own books. If the UFO evidence was confusing by scientific standards, the supposed evidence of ancient visitations was mostly embarrassing. Although the idea had been broached before and would become a publishing phenomenon later, it was the German writer Erich von Däniken who made ancient astronauts a popular theme worldwide during the 1970s. As the Condon committee was finishing its report, von Däniken was writing his *Chariots of the Gods?* (1970), followed by many more books in the 1970s. The question of ancient astronauts was not invalid; given the time scale of the universe, it is not out of the question that Earth could have been visited in the past. The challenge was to come up with substantial evidence, and here almost all scientists, and to its credit even much of the public, found von Däniken wanting.

The controversy over crashed spaceships, alien contactees, and abductees was even worse from the point of view of most reputable scientists. Books such as *The Roswell Incident* (1980), *UFO Crash at Roswell* (1991), and *Crash at Corona* (1997), purporting to describe a saucer crash in 1947, simply could not be taken seriously by most scientists. A U.S. Air Force report released in 1994, concluding that the Roswell evidence was part of the debris of a previously secret "project MOGUL" designed to search the high atmosphere for reverberations from Soviet nuclear tests, did nothing to dampen the enthusiasm of most Roswell believers for the alien hypothesis. A final Air Force study, *The Roswell Report: Case Closed* (1997), detailed how anthropomorphic test dummies might have been mistaken for alien bodies; this again brought more skepticism. Tens of thousands showed up in Roswell in 1997 for the 50th anniversary of the supposed saucer crash, an event eerily juxtaposed with coverage of Pathfinder's landing on Mars on July 4. Roswell, "Area 51," and other supposed alien body repositories had become a part of American folklore, glamorized in movies such as *Independence Day* and testimony to American skepticism of government. But they had not become a part of American science.

As for contactees and abductees, what in the 1950s and 1960s had been a minor diversion to the UFO theme with the claims of Adamski and the Hills, came in the 1970s and 1980s to dominate it. Two fishermen in Mississippi claimed to have been given a physical exam aboard a flying saucer in 1973. Budd Hopkins documented his abduction claims in *Missing Time* (1981) and *Intruders* (1987), the latter published by the prestigious Random House. Whitley Strieber's claims of abduction in *Communion: A True Story* (1987) spawned a best-seller. In *Secret Life: Firsthand Accounts of UFO Abductions* (1992), the UFO historian David Jacobs argued the case for alien abductions, based largely on statements of subjects under hypnosis. In the same year an Abduction Study Conference was held at MIT, documented in detail in C. D. B. Bryan's *Close Encounters of the Fourth Kind* (1995). And in 1994, the work of Hopkins, Jacobs, and others was given a further boost when the formerly skeptical Harvard psychiatrist John Mack concluded in *Abductions: Human Encounters with Aliens* that the phenomenon could not be explained psychiatrically, was not possible within the framework of the modern scientific worldview, and was in all likelihood truly explained by alien abduction. All of these claims were amplified by radio and television talk shows and even news reports. Once again, in a repeat of the 1950s UFO phenomenon, most scientists were only too happy to leave this discussion to the media. The phenomenal popularity of the subject demonstrated the continued willingness of the public to accept, without physical evidence, even the most extreme beliefs of the extraterrestrial hypothesis – close encounters of the third kind – or at least to use them for entertainment value. In short, such beliefs demonstrated allegiance to a different worldview than the scientific, as normally understood.

Lost in a sea of paranormal claims that brought it into disrepute among scientists, the extraterrestrial hypothesis after Condon swung back toward the pre-1965 media excesses, becoming virtually the exclusive property of authors with mixed motives and unscientific credentials. As such claims spread, the UFO phenomenon clearly fell from the province of scientific discourse. The hopes of those who wished to expand the scope of science were largely destroyed by the absurd claims of those whose motives were closer to profit than truth. The signal-to-noise ratio problem that had plagued the UFO phenomenon from the beginning was now joined by the signal-to-noise problem of unscrupulous authors.

By the end of the century the question was not whether extraterrestrials were visiting Earth, but whether there was anything at all to the UFO phenomenon that science could productively illuminate. Neither normal scientific method nor the norms of interaction among scientists held out much hope. Gerard Kuiper believed that scientists did not avoid the UFO question from fear of ridicule, but rather because it was a question unlikely to be answered with the available evidence. This may well underestimate the role of scientific peer pressure, which, combined with the elusiveness of the problem and the drift toward pseudoscience, is likely to render further research slow at best. The continued interest of scientists in the aftermath of the Condon study is underscored in a poll of astronomers' attitudes toward UFOs undertaken by Sturrock in 1977, the first since Hynek's 25 years before. It showed that 53 percent of responding astronomers believed the UFO problem deserved further study. Sixty-two respondents even claimed they had seen or obtained an instrumental record of an event they believed to be related to UFOs. But only 13 percent of all respondents could see any way to solve the problem of identifying UFOs. If an objective of science was "a consensus of rational opinion," Sturrock concluded, it had failed in the case of UFOs.

Thus, in addition to career considerations and peer pressure, the relative inaction of scientists by the end of the century seemed an admission that a solution to the problem might have to await further developments in science. While the UFO phenomenon may be beyond the understanding of 20th-century physics, Hynek liked to say, "there will surely be, we hope, a twenty-first century science and a thirtieth century science, and perhaps they will encompass the UFO phenomenon as twentieth century science has encompassed the aurora borealis, a feat unimaginable to nineteenth century science, which likewise was incapable of explaining how the sun and stars shine." Meanwhile, the history of the debate finds abundant blame on all sides: the unwillingness of many scientists to actively engage the UFO phenomenon is certainly understandable in terms of peer pressure and career advancement, but the desire of some to squash the subject without considering the evidence flies in the face of scientific curiosity that supposedly led them to science to begin with. On the other side, the outrageous claims and hoaxes that presently flood the field are unworthy of scientific attention, and no one should be surprised if scientists fail to engage every will-of-the-wisp report. In the middle of these extremes may yet be a phenomenon that requires study, if only one can find it in the midst of the twin human failings of perception and deception.

In the meantime, lacking any definitive resolution, at the very least the UFO phenomenon holds an important place in the history of the mythic imagination. Increasingly, toward the end of the century, some scholars and scientists began to view it in just this way. One study called the phenomenon "the first myth to develop in the modern, high-tech, instant global communications world." Another asked, "Do fairies, ghosts, and extraterrestrials exist as living beings – or are they some product of the human mind? The obvious answer to which the evidence overwhelmingly, unemotionally and logically points is a resounding no! They are mental constructs." Even Vallee admitted that extraterrestrials could be viewed as a modern version of demons and elves.

Whether or not any definitive explanation of the UFO phenomenon is forthcoming, historically the effect of the extraterrestrial hypothesis of UFOs on the extraterrestrial life debate was multifaceted. There is no doubt that it brought scientists out of the closet on a subject they otherwise might never have addressed. It was the public's chief exposure to the subject of extraterrestrial life, and even as the more sober and scientific SETI programs were undertaken, UFOs remained prominent in the public mind. For scientists the decline of the extraterrestrial hypothesis of UFOs did not mean the decline of belief in extraterrestrial life; even Menzel had always kept the two distinct. Even so, it is remarkable that proponents of extraterrestrial intelligence would be able to push ahead with their SETI programs even as the UFO phenomenon brought the subject of extraterrestrials on Earth into disrepute.

The attempts of scientists to understand the UFO phenomenon have close parallels with the rest of the extraterrestrial life debate. We recall (Chapter 3) Peter van de Kamp's comment on the extreme difficulty of observing extrasolar planetary systems: "Should we not come to the rescue of a cosmic phenomenon trying to reveal itself in a sea of errors?" he asked. Though van de Kamp was dealing with instrumental measurements, and though the word "cosmic" may prove to be irrelevant to the UFO debate, the phenomenon has surely been plagued by a sea of errors no less than other, more prosaic attempts to find life beyond the Earth. Whether or not a true natural phenomenon is trying to reveal itself, only the future will tell. But if and when extraterrestrial intelligence is discovered in the distant reaches of space, it will surely be recalled that for a time in the 20th century, more than a few Earthlings believed the extraterrestrials had actually frequented the skies of their home planet, crashed or landed, and abducted millions of its citizens.

6

THE ORIGIN AND EVOLUTION
OF LIFE IN THE
EXTRATERRESTRIAL CONTEXT

*There is every reason now to see in the origin of life not a "happy accident"
but a completely regular phenomenon, an inherent component of the total evolutionary development of our planet. The search for life beyond Earth is thus
only a part of the more general question which confronts science, of the origin
of life in the universe.*

A. I. Oparin (1975)

*A full realization of the near impossibility of an origin of life brings home the
point how improbable this event was. This is why so many biologists believe
that the origin of life was a unique event. The chances that this improbable
phenomenon could have occurred several times is exceedingly small, no matter
how many millions of planets in the universe.*

Ernst Mayr (1982)

*. . . Nowhere in all space or on a thousand worlds will there be men to share
our loneliness. There may be wisdom; there may be power; somewhere across
space great instruments, handled by strange, manipulative organs, may stare
vainly at our floating cloud wrack, their owners yearning as we yearn. Nevertheless, in the nature of life and in the principles of evolution we have had our
answer. Of men elsewhere, and beyond, there will be none forever.*

Loren Eiseley (1957)

*From the perspective of determinism and constrained contingency that pervades
the history of life . . . life and mind emerge not as the results of freakish acci-
dents, but as natural manifestations of matter, written into the fabric of the
universe.*

Christian de Duve (1995)

Although the origin of life is today viewed as central to the question of life in
the universe, prior to the middle of the 20th century, theories of the origin and
evolution of life played very little role in the extraterrestrial life debate. The
reason is in part that extraterrestrial life had historically been the province
of astronomers rather than biologists. The determination of conditions on
the planets and the search for planets around other stars were physical questions logically prior to any discussion of biological details. And when it came
to biology, astronomers for the most part simply assumed that since life had
originated on Earth, it would originate on any other planet under proper
conditions. Percival Lowell and other astronomers were therefore content to
determine planetary conditions rather than to study mechanisms of how life
might have originated.

But the lack of biological input into the extraterrestrial life debate in the first half of the century was also a reflection of the state of biology itself. The subject of extraterrestrial life was simply beyond the concern of most biologists during this period for the very good reason that biology itself was not yet a mature discipline. Modern biology as an autonomous science is generally considered to have begun only with the events precipitated by Darwin's *Origin of Species* in 1859, and not until a century later were the biological sciences unified via the evolutionary synthesis. It is therefore not surprising, from the point of view of the maturity of biology, that its studies were largely confined to the Earth's biosphere.

Not only was biology still immature during the first half of the 20th century, it was also inferior to physical science in the sense that, unlike Newtonian physics, it lacked universality. Confined to the Earth, no "biological law," however unified, could compete for status with the universality of Newton's laws. Just at the time the unification of the biological sciences was complete, however, the Space Age brought the possibility of extending biology beyond the Earth and, by searching for life on Mars (and eventually beyond the solar system), aspiring to the same status of universality held by the physical sciences. The impressive progress in the study of terrestrial biology did not allow one to separate the contingent from the necessary in living systems, including its chemical basis, the role of proteins and nucleic acids, and the probability of life's origin in the first place. "To the extent that we cannot answer these questions," wrote a panel of top biologists commissioned by the U.S. National Academy of Sciences to consider the problem of life on Mars in the early 1960s, "we lack a true theoretical biology as against an elaborate natural history of life on this planet. . . . The existence and accessibility of Martian life would mark the beginning of a true general biology, of which the terrestrial is a special case." This search for a "true general biology" was the promise, and the hope, of exobiology, and we shall see it embodied in this chapter in the search for the origins and evolution of life.

6.1 ORIGINS OF LIFE AND EXTRATERRESTRIAL LIFE: A SPACE AGE SYMBIOSIS

Without doubt the Space Age was the single most important driving force behind the marriage between biology and astronomy in the second half of the 20th century. It is true that early in the century the Swedish chemist and physicist Svante Arrhenius had championed a "panspermia" theory, whereby the seeds of life were spread from planet to planet and star to star by radiation pressure; this theory, however, was never widely accepted and had been dead since the 1920s. The chemical theory of the origin of life, expounded independently in the 1920s by the Russian biochemist A. I. Oparin and the

British biochemist J. B. S. Haldane, proposed that the problem of the origin of life should be explained entirely in terms of the general laws of physics and chemistry as applied to primitive Earth conditions. Oparin and Haldane shared the basic precepts that the early Earth atmosphere was hydrogen rich ("reducing"), that organic substances would be synthesized in such an atmosphere, and that these organics accumulated until the primitive ocean (in Haldane's words) "reached the consistency of hot dilute soup." Moreover, they agreed independently that these organics would form larger and more complex molecules, resulting in "the first living or half living things" such as viruses, and eventually in unicellular organisms such as bacteria. In short, despite some differences, they agreed that life could arise from nonlife, and their theory became known as the "Oparin–Haldane theory." Although it was clear that this theory could be applied to life throughout the universe, for decades it remained almost entirely Earthbound.

The extraterrestrial implications of the Oparin–Haldane theory remained largely latent even when the famous Urey–Miller experiment in 1953 produced amino acids from a mixture of chemicals believed to have been present on the primitive Earth. But the probability that similar events would have occurred anywhere in the universe under similar conditions could not be denied, a probability that was pointed out by the Harvard biologist George Wald in a *Scientific American* article on "The Origin of Life" as early as 1954 and in the same year by Haldane himself in the context of the coming Space Age. Not only did the years immediately following 1957 see the rapid application of origin of life studies to problems raised by the Space Age, and the application of Space Age data to origin of life studies, modified panspermia theories also gradually came back into vogue at this time. In short, the Space Age saw the integration of two subjects that had been almost entirely separate since the Arrhenius hypothesis at the beginning of the century.

Within the context of the Space Age, one may distinguish at least four interrelated ways in which this integration of origin of life and extraterrestrial life studies occurred. First, biological scientists (including primarily biochemists and geneticists) were naturally drawn into such pressing Space Age issues as planetary contamination and life detection, issues that suddenly became national concerns. Intellectually they saw that these issues held the potential to contribute to long-standing biological problems, and the considerable NASA funding suddenly available for biological research was for many a novel, and difficult-to-refuse, proposition. For the first time, biologists and space scientists worked together toward a common goal. In the process, NASA transported origin of life studies – and their accompanying biologists – into outer space. Second, beyond those specific Space Age goals, theories and experiments in the origin of life were increasingly seen in an extraterrestrial context, whether or not they were funded by NASA or driven by specific Space Age goals. Origin of life was one of the few areas in which biological science could aspire to universality. That aspiration did not begin or end with Viking but had an internal dynamic of its own, felt in varying degrees by origin of life practitioners. Continuing a process initiated by the Space Age, conferences on the origin of life, previously confined to terrestrial concerns, expanded to the realm of the extraterrestrial, while conferences on extraterrestrial life, traditionally largely astronomical, included an increasingly biological component. Third, even had the aspiration for a universal biology not been present, the discovery of amino acids in carbonaceous meteorites, and of complex organic molecules in interstellar molecular clouds, comets, and interplanetary dust (not to mention the famous Martian meteorite) forced biological interest in the extraterrestrial realm and sustained it even after the issues of planetary contamination and life detection were resolved (to the satisfaction of most) by the Viking landers. Finally, philosophical issues such as chance, necessity, and the nature of life, endemic to the origin of terrestrial life, were not only shared but even more crucial in the extraterrestrial realm. We shall review in turn the roles of each of these four factors.

Space Age Issues

As we have seen in Chapter 2 in the context of solar system studies, the Space Age, with its related concerns for the prevention of contamination of the extraterrestrial environment and its hopes for the detection of life beyond the Earth, raised immediate substantive issues of interest to those studying the origin of life. The contamination issue originally emanated from the biological side of the National Academy of Sciences rather than from NASA, and from the beginning it naturally drew on experts from the biological sciences. We recall that it was the National Academy that sponsored two committees, one headed by the geneticist Joshua Lederberg and the other by the biochemist Melvin Calvin, to study the problem of extraterrestrial life, with special emphasis on contamination. It was NASA, of course, that led the way for programs to detect life on Mars, eventually by necessity becoming deeply involved with the contamination issue as well. Nor were these interests exclusively American. If the extraterrestrial environment were going to be contaminated, it could just as well happen via Russian microbes as American ones – a terrible irony considering Oparin's work. Similarly, Russian spacecraft could just as well be the first to discover life on Mars; even though those spacecraft missed Mars and in any case carried no sophisticated life detection experiments, the issues were discussed in the Soviet Union as well as in the United States. These extraterrestrial components of the origin of life thus had international ramifications, even if only in a bipolar sense in terms of the space powers.

Fig. 6.1. Organic chemist Melvin Calvin, pioneer in photosynthesis and origin of life research, Nobel Prize winner, and exobiology enthusiast. Graphic Arts Department, University of California, Lawrence Berkeley Laboratory.

As one example of how the biological community became involved with the issue of extraterrestrial life via NASA, we may trace the entry into the field of Melvin Calvin (Fig. 6.1), the pioneer in photosynthesis and origin of life research, who, according to the first head of NASA's exobiology office, "did more than anyone else to establish the posture of the nation in bracing itself for the search for extraterrestrial life." Calvin (1911–1997), inspired by George Gaylord Simpson's *The Meaning of Evolution* (1949), had been working on the problem of chemical evolution since 1950 at the University of California, Berkeley, and had done experiments on prebiotic evolution using hydrogen, carbon dioxide, and water. Unlike Urey and Miller, who 3 years later performed their famous experiment using methane and ammonia in a reducing atmosphere, Calvin did not produce amino acids, leaving his experiments less well known (at least to the public) than those that produced organic components. His fame had to await his work in another related area – photosynthesis – for which he received the Nobel Prize in 1961. But, Calvin recalled many years later, it was while performing the experiments on prebiotic evolution in about 1950 that he came to the private conclusion that "there must have been evolutionary processes of the same sort elsewhere in the solar system, and perhaps elsewhere in the universe." The reason for this belief was that "there was no question in my mind about the Darwinian behavior of molecules. And if that were true, then that meant it must be true in the entire universe."

Although the extraterrestrial component does not appear in Calvin's lectures on the origin of life in 1956, by late 1958 Calvin was chairing the Panel on Extraterrestrial Life to consider problems of contamination by spacecraft. And by early 1959 he was participating in the Lunar and Planetary Exploration Colloquia being held in California. It was here that he gave his first detailed statement on the relation of origin of life research to extraterrestrial environments. After discussing in detail the latest research on the origin of life, Calvin concluded that in view of what was known about the origin of life on Earth, all that was needed to determine the prevalence of life in the universe was an estimate of the number of sites with conditions similar to those on Earth. Such estimates Calvin found in the works of the astronomers Fred Hoyle and Harlow Shapley, adopting the latter's estimate of 100 million suitable planets in the universe. Like Wald and others before him, Calvin assumed that when conditions were present, life would evolve, a premise he carried even to the case of advanced intelligence.

Calvin stands as only one example of how a biochemist entered the realm of the extraterrestrial: experiments in photosynthesis and chemical evolution led to an interest in the origin of life; the belief that chemical evolution was occurring beyond the Earth led to the conviction that life also must be originating there; and the origin of life on planets such as Mars became a prime goal of the Space Age, allowing Calvin not only to explore his beliefs, but also to bring along many others as well. Each biologist would travel his own unique career path to outer space. Whether these were forward-looking geneticists such as Lederberg or biologically inclined astronomers such as Carl Sagan, the Space Age allowed each to bring his interests to bear on the problem of extraterrestrial life and a select few to participate on the Viking science teams. Key biologists perceived the mutual benefits of their involvement in matters extraterrestrial, and not only inserted themselves into appropriate committees, but also began to bring along some of their biological colleagues.

NASA further encouraged biological involvement by establishing not only ad hoc panels of consultants, but an entire life sciences program. When in

1960 the space agency began that life sciences program, it became an important center of activity for origin of life studies, particularly as applicable to the space program. In July 1959 Administrator T. Keith Glennan formed a Biosciences Advisory Committee, chaired by Dr. Seymour Kety, to examine NASA's long-term interests in the life sciences. Their January 1960 report endorsed activities far beyond the immediate operational interests of the space program, including "basic biologic effects of extraterrestrial environments . . . and identification of complex organic or other molecules in planetary atmospheres which might be precursors or evidence of extraterrestrial life." It was during the early formation of NASA's exobiology program that Calvin was drawn into NASA's activities as chairman of the Space Biology Committee.

Among the recommendations of the Kety committee was the establishment of a NASA life sciences research facility, which was eventually won by Ames Research Center near San Francisco. Here NASA set up its exobiology program, first under Harold P. "Chuck" Klein (later of Viking fame), then under Richard S. Young when Klein became head of the life sciences program at Ames. Cyril Ponnamperuma (a student of Calvin) came in as a postdoctoral associate and ultimately headed a branch in chemical evolution: experiments on origin of life were an important part of the research. By the mid-1960s, origin of life studies were not only strongly integrated into the space program, they were essential to one of its most highly visible programs – the search for life on Mars. The work of Ponnamperuma and others at NASA's life sciences laboratory was not pure research, but related to spaceflight goals. When Miller and the Caltech biochemist Norman Horowitz reviewed experimental progress in the origin of life for the Space Studies Board of the National Academy of Sciences, it was no mere abstract exercise, but an essential task with the end in view that the ideas "were important for the design of experiments to detect life on Mars." The same pattern was repeated again and again during the 1960s, and the culmination in the Viking program is well known. Because of its decision to search for life on Mars, NASA nourished exobiology, and a preeminent part of exobiology was the study of the origin of life.

At the same time, it is clear that biologists had their own agenda. Aside from urgent issues such as planetary contamination and life detection, the biological community was quick to see the potential benefits of the space program to its own work, including specifically origin of life issues, which raised the possibility of a generalized biology. Already in 1959, Stanley Miller and Harold Urey had given a justification for both exobiology and origin of life studies when they concluded in another of their papers on organic synthesis that "Surely one of the most marvelous feats of the 20th century would be the firm proof that life exists on another planet. All the projected space flights and the high costs of such developments would be fully justified if they were able to establish the existence of life on either Mars or Venus. In that case, the thesis that life develops spontaneously when the conditions are favorable would be far more firmly established, and our whole view of the problem of the origin of life would be confirmed." Thus, barely 6 months after the formation of NASA, at least some forward-looking biologists realized they were literally entering a new world. Biochemists such as Calvin and Horowitz were not only essential to NASA, they were also extremely interested in how the search for extraterrestrial life could inform their own research.

Aspirations for Universality

Even as the space program catalyzed the integration of origin of life and extraterrestrial life studies, theories and experiments previously viewed in the context of life's beginning on Earth were increasingly seen in an extraterrestrial context, raising the two fields of study to a more permanent level of integration. In the Soviet Union, Oparin, himself not involved in the Soviet or American search for life on Mars, wrote at length on the origin of life in space in 1964, and a few years later he called exobiology and the origin of life "the inseparable connection." Noting that the origin of life on Earth was no longer seen as a happy accident, but rather as a normal evolutionary process following chemical evolution, Oparin held that "Study of the conception of life on Earth amounts to an investigation of only one example of an event which must have occurred countless times in the world. Therefore, an explanation of how life appeared on Earth should strongly support the theory of existence of life on other bodies in the universe." After the mid-1960s, treatises on the origin of life almost always encompassed extraterrestrial life, and treatises on exobiology had the origin of life as one of their principal foundations. And by 1970 it was no surprise at all that a National Academy committee surveying the life sciences should include in its study *Biology and the Future of Man* a substantial discussion of extraterrestrial life in its chapter on origin of life; indeed, its absence would have been considered a serious omission.

While research on the synthesis of organic compounds and theories of the origin of life could proceed without regard to the question of extraterrestrial life, at a deeper conceptual level the two were becoming increasingly intertwined; biology was pursuing unification and universality at the same time. Nowhere is the nature of the integration more evident than in the volume *Exobiology* (1972) by Cyril Ponnamperuma, a student of Calvin involved with NASA biology from early in his career. *Exobiology* was one of a series of volumes in which the general editors proclaimed that "the sharp boundaries between the various classical biological disciplines are rapidly disappearing" and that biological specialists needed to keep abreast of a broad number of fields; it is significant that by 1972 exobiology was viewed as one of the "frontiers of biology" that needed to be thus absorbed. Laboratory experiments on the origin of life, Ponnamperuma wrote, were intended to help us understand not only life on Earth, but the origin of life in the universe. Thus, the major portion of the volume reported on ideas and experiments undertaken as the basis for understanding the origin of both terrestrial and extraterrestrial life. Such an integration became common practice, and it was strengthened by the observation of organic molecules in meteorites and other outer-space environments.

Great difficulties, however, remained even in understanding the origin of life on Earth. Despite the great promise of prebiotic synthesis beginning with the experiments of the 1950s, by the end of the century most of the steps to life had not been duplicated in the laboratory under conditions believed to exist on the primitive Earth. Just how far laboratory syntheses remained from actually producing life is clear from Figure 6.2, which shows in general outline major concepts in the evolution of chemical and biological complexity, including the uncertainties in the relative roles of proteins and nucleic acids at the interface between chemical and biological evolution. Following the Miller–Urey experiments in 1953, it became clear that amino acids were not difficult to synthesize; by early 1970, 17 of the 20 amino acids occurring in proteins had been synthesized under possible primitive Earth conditions. Yet, 25 years later, no Miller–Urey experiment had produced the much more complex proteins, nucleic acids, polysaccharides, or lipids that make up a cell, although some had been synthesized under non-Earth-like conditions. In fact, none of the nucleotides or nucleosides that compose the nucleic acids had been synthesized under such conditions. In this respect the program of chemical evolution was a disappointment.

More fundamentally, as the biochemist Robert Shapiro pointed out in a critical analysis of origin of life theories in 1987, three out of four of the parts of the mature Oparin–Haldane hypothesis – the "reducing" atmosphere of hydrogen compounds, the accumulated hot dilute soup of organic chemicals, and the development of life out of this soup – lacked confirmation. In particular, the assumption of a strongly reducing primitive atmosphere grew weaker with new geochemical evidence, bringing into question the relevance of all the Miller–Urey experiments. Given these uncertainties, other alternatives were proposed to make the origin of life less problematic: the astronomically ubiquitous compound hydrogen cyanide (HCN) might polymerize nucleosides and nucleotides directly rather than via amino acids; further up Figure 6.2, RNA might be more easily synthesized than DNA or proteins as the first replicative molecule; even more radically, perhaps life had originated with clay minerals before the "genetic takeover" by organic molecules inaugurated biological evolution as we know it. Although many were willing to rethink the reducing atmosphere of the Oparin–Haldane theory along with

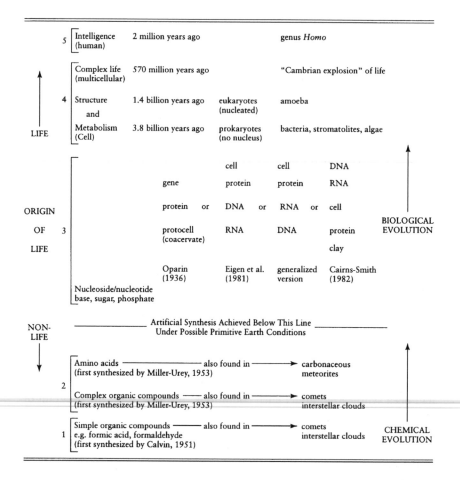

Fig. 6.2. Evolution of chemical and biological complexity. At levels 1 and 2 artificial synthesis is easily achieved under possible primitive Earth conditions. Synthesis has also been achieved for some components at level 3, but not under possible Earth-like conditions. At level 3 it is uncertain whether RNA, DNA, proteins, or the cell framework came first, though Oparin's proposed sequence dominated for 40 years. The "gene" of Oparin's theory was known to be composed of RNA and DNA after 1944. Life has been described as beginning at level 3 with the proposed RNA "random replicator" of Eigen's theory, or the clay crystal directing the synthesis of protein, or at level 4 with the cell. At level 5 intelligence evolved gradually and is not easily defined, but the human brain has 16 orders of magnitude more atoms than the cell.

other of its aspects, most were not willing to go so far as to abandon carbon as the basis for life's origin.

While not all origin of life practitioners saw their work in an extraterrestrial context, the relevance of these theories and experiments to extraterrestrial life is clear from many of the participants, quite aside from the general claims of connection stated by Oparin, Ponnamperuma, and others. Moreover, conferences on the origin of life and on extraterrestrial life increasingly sought to establish the mutual relevance of the two domains, precipitating not only a merging of concepts but also a mingling of researchers in the two fields. Whereas extraterrestrial life played only a very minor role in the first conference on origin of life in Moscow in 1957, by the second international meeting in 1963 there was considerably more impetus from the problem of extraterrestrial life. Indeed, NASA was one of the sponsors of the meeting, and it was attended by many who now undertook their experiments with NASA funding. In 1968 NASA sponsored its own conference on the subject, and at subsequent international meetings (usually held every 3 or 4 years), extraterrestrial considerations became a significant aspect of the discussions.

Similarly, conferences on extraterrestrial life drew on origin of life experts from the start. As we shall see in the next chapter, Melvin Calvin played a significant role at the first conference on communication with extraterrestrial intelligence, held at Green Bank, West Virginia, in 1961. At the JPL meeting on "Current Research in Exobiology" in 1963, the school of chemical synthesis of organic molecules was represented by Juan Oró and the meteorite controversy, simulated extraterrestrial environments, contamination, and life detection experiments each by its own experts. Since then, virtually every conference on the general problems of extraterrestrial life has included origin of life as a major component. Both in the transfer of concepts and in the mingling of the two scientific communities, such conferences were an essential glue in the integration of origin of life and extraterrestrial life studies. Via origin of life, extraterrestrial life was the biological sciences' window on its aspirations for universality.

Revival of Panspermia

Yet another pathway to integration was the revival of panspermia in many forms. Just as Arrhenius turned to panspermic theories of life from outer space after spontaneous generation was disproved, the increasing uncertainties of Earthbound prebiotic synthesis led some to turn once again to outer space. Although Carl Sagan emphasized the difficulties of the classical panspermia idea in light of modern astronomy, the trend was toward a neo-panspermia radically different from Arrhenius's version. The central focus of the new panspermia in the last four decades of the 20th century was not life itself drifting through space, but prebiotic chemicals. Observational evidence of organic synthesis in meteorites, comets, dust, ice, the atmospheres of the Jovian planets or their satellites, and interstellar molecular clouds showed nature to be proficient at building the chemistry of life. Such observations raised the possibility that organic synthesis, which Oparin and Haldane postulated had occurred in the Earth's primitive atmosphere, might have taken place even before the formation of the Earth and been delivered later by impacts from outer space. Already in the early 1960s, as Oró and others gave birth to the field of organic cosmochemistry, Bernal had made just this suggestion. The discovery that life on Earth may have begun 3.8 billion years ago, just at the end of heavy bombardment by cometary and meteoritic material, raised interest to new heights.

Once again, however, the interpretations of these observations were ambiguous. Whether, and if so just how, organic synthesis in space was related to the origin of life on Earth were open questions. So was its lesson for the existence of life on other planets. There was no doubt that the detection of organic molecules in interstellar space and in a variety of interstellar and interplanetary environments were first-rate scientific discoveries. What is particularly interesting in the present context, however, is the use made of these discoveries in the extraterrestrial life debate. In short, while the existence of organics in space was firmly established, their lesson for life would become a subject of considerable debate. Exobiologists lost no opportunity to point out that the ease with which organics were apparently formed boded well for extraterrestrial life; critics were quick to emphasize the considerable distance between even complex organic molecules and life.

The first claims of organic matter from space came from the study of meteorites, a connection made already in the 1870s and 1880s, with another brief spurt of activity in the 1930s. That certain stony meteorites contained carbon was known from the work of the Swedish chemist J. J. Berzelius in 1834, but it would turn out that these "carbonaceous chondrites" were the rarest form of meteorite. Only in the 1960s did analytic techniques allow the nature of the more complex compounds to be determined. Once again, Melvin Calvin and his colleagues were ahead of their time. Having obtained from the Smithsonian Institution a piece of the carbonaceous chondrite that had fallen near Murray, Kentucky, in 1950, in the opening days of the 1960s they reported at the First International Space Science Symposium in France "the presence in meteorites of complex organic materials, some of them apparently uniquely pertinent to life processes." Among the molecules reported was a chemical similar to cytosine, one of the four bases in the DNA molecule. Notably absent were any traces of amino acids. The lesson Calvin drew from the analysis was that the question "as to whether or not there were possibly prebiotic forms out on astral bodies other than the earth, seems to be answered, at least tentatively, in the affirmative."

Following this claim, at a meeting of the New York Academy of Sciences in March 1961, the team of Bartholomew Nagy, Warren Meinschein, and Douglass Hennessy reported that material from the Orgueil carbonaceous meteorite, which fell near Orgeuil, France, in 1864, included "paraffinoid hydrocarbons" characteristic of living organisms. Based on the similarity of the detected distribution of hydrocarbons to that found in animal products such as butter, the authors concluded that they represented biogenic activity, and they further inferred that "biogenic processes occur and that living forms exist in regions of the universe beyond the earth." Bernal was quick to point out the consequences of this claim. Writing in *Nature* 3 weeks later, he urged that scientists not ignore these analyses just because of widespread media publicity; "whatever the interpretation put on them they are of cardinal importance to science." The evidence, Bernal noted, allowed only "that the meteorite material may be of organismal origin and not that it must be so." In other words, it was possible that the carbonaceous material was formed by some form of life or by wholly inorganic processes. If the former, the material must have been produced by life on the planetary body from which the meteorite arose. Furthermore, Bernal suggested that meteorites could be the source of material for the first synthesis of life on Earth.

Even as this controversy was heating up, Nagy and George Claus, a microbiologist at New York University Medical Center, made a further startling claim of quite a different kind. They asserted that two of the meteorite samples they examined (the Orgueil and Ivuna) contained five types of "organized elements," structures they identified as possible remnants of organisms "resembling fossil algae" (Fig. 6.3). This was an interpretation disputed by most; intricately patterned hardly meant alive. And Edward Anders and Frank Fitch at the University of Chicago showed that at least some of the results were likely due to contaminants. A raft of papers in the March 1962 issue of *Nature* showed just how difficult were the interpretations of these photographs. Bernal supported the microfossil theory, and although Urey was more skeptical, he supported further work. Although by 1975 Nagy himself had come to the conclusion that a biological interpretation of the so-called organized elements in meteorites was only a "remote possibility," and although a consensus developed that the phenomenon was probably due to degraded terrestrial pollen, the variety of conclusions demonstrated that for the origin of life, interpretation was problematic even when the object was at hand and subject to experiment. By the end of the century, especially after the findings in the Mars rock, reputable researchers were suggesting a reexamination of the Claus and Nagy work.

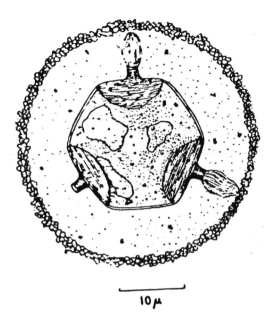

10μ

Fig. 6.3. Sketch of an organized element found in carbonaceous meteorites, believed by Nagy and Claus to be possible remnants of living organisms. In some ways, the controversy foreshadowed the Martian fossil controversy of the late 1990s. Reprinted with permission from Claus and Nagy, *Nature* (1961). Copyright 1961 Macmillan Magazines Limited.

Despite the apparent lack of proven biogenic "organized elements" in carbonaceous meteorites, the early 1970s advanced spectacularly beyond the claims of simple organic content. The fall of a carbonaceous chondrite near Murchison, Australia, in 1969 provided the opportunity to analyze the meteorite with a minimum of contamination concerns, one of the primary ambiguities in previous experiments. A sample analyzed at NASA's Ames Research Center – still NASA's chief life sciences laboratory – gave unambiguous results: 74 amino acids, the primary building blocks of life, 55 of which did not occur on Earth, including "right-handed" forms not found in terrestrial living systems. Subsequent confirmation in other carbonaceous meteorites, including many relatively uncontaminated in the highlands of the Antarctic, led many to believe that this was their standard composition, and that others in which it was not found had been altered with the passage of time. Although most believed the amino acids in carbonaceous meteorites to be cosmochemical (originating in the solar nebula, for example) and not biogenic, the fact that they have been generated at all beyond the Earth was seen by many as further evidence of the possibilities of life originating in extraterrestrial environments. And as we have seen in Chapter 2, the announcement in 1996 not only of organics but also of possible fossil life in the Martian meteorite ALH 84001 raised interest in meteorites and panspermia to new heights. Indeed, by 1997 there were claims of microfossils in the Murchison meteorite.

Even as the controversy raged over meteorites, comets had long been known from optical observations to possess simple carbon-bearing molecules such as C_2, C_3, and CH. By 1974 observations of comet Kohoutek had brought the first definitive detection of the organic molecules HCN and methyl cyanide (CH_3CN). While Oró and others continually stressed the known existence of organics in comets, and the possible existence of biochemical compounds such as amino acids and purines, they also emphasized that it was unlikely that chemical evolution had progressed beyond that seen in carbonaceous chondritic meteorites. Just how cometary organic materials affected the origin of life on Earth remained problematic. Aside from a direct impact such as may have occurred in Siberia in 1908, some suggested that they might have entered the Earth in the form of interplanetary dust particles that originated with comets. The survival of any organic material hitting the Earth at high velocities, however, was in question. Others were less cautious, including the astronomer Fred Hoyle and his colleague Chandra Wickramasinghe, who claimed that the organic material in comets might actually be in the form of bacteria.

By the late 1960s the first organic molecules, including ammonia (NH_3), formaldehyde (H_2CO), and HCN, had also been found in molecular clouds in outer space. Three decades later, some 65 of the more than 100 molecular species discovered in space had been identified as organic, as evidenced mainly by their characteristic radio signature. Complex organic molecules including polycyclic aromatic hydrocarbons (PAHs) had also been found by infrared observations of interstellar ice and dust. In whatever form they occurred, their discoverers wrestled with the question of the relation of these organic molecules to life: "The predominance of organic species and their similarity to the products obtained in the synthesis of amino-acids in the study of the origin of life suggests a very close parallel between interstellar clouds and prebiotic chemistry," the astronomer David Buhl noted in reviewing the early discoveries in 1971. However, he was quick to add that "It is difficult to establish a direct connexion, but the similarity of the organic molecules produced in laboratory experiments to the molecular composition of interstellar clouds suggests a dominant direction in the chemical synthesis which proceeds despite the very different environment of the cool, low density interstellar clouds." Carl Sagan came to the same conclusion in the early 1970s, but by the early 1990s he argued with his colleague Christopher Chyba that the delivery of organics from outer space might well have played a role in the origin of life.

Meanwhile, in our own solar system, evidence had been accumulating of organics in some of the giant gaseous planets and their satellites. As early as his classic book *The Planets* (1952), Urey had suggested that organic molecules might cause Jupiter's multicolored appearance. By 1971 Sagan concluded in a review article not only that "it seems rather clear that very large quantities of organic molecules must exist on Jupiter today, that quite complex organic reaction chains are operating, and that organic molecules are an important presumptive source of the Jovian coloration," but also that "there appear to be some interesting exobiological opportunities on Io, Europa, Ganymede, Callisto, Titan, Triton and Pluto." Even after the flight of the Voyager spacecraft, however, the complexity of Jupiter's organics, and the issue of whether organic or inorganic chemistry caused its coloration, were still in question. Attention increasingly focused on Saturn's satellite Titan, which Voyager's infrared spectrometer showed to have a rich assortment of at least nine simple organic molecules ranging from hydrocarbons to HCN. As in the case of Jupiter, the level of complexity of organics remained unknown, but Sagan and his colleagues led the way in laboratory experiments of simulated Titan atmospheres showing that complex organic solids known as "tholins" were the likely candidate for causing the reddish color of Titan's atmosphere. This was one of the pressing problems to be addressed as the Cassini spacecraft headed toward Saturn at the end of the century. Meanwhile, the Galileo spacecraft's observations increased the likelihood of an ocean with possible organics on the Jovian moon Europa. Ironically, it turned out that more organics existed beyond Mars than on Mars, the planet on which a billion dollars had been spent to search for life.

By the mid-1970s, then, evidence of amino acids in meteorites, and of lesser organics in comets and interstellar clouds, had been well established, while the existence of at least simple organics in the atmospheres of some of the Jovian planets and satellites was soon to be proved. Although the discovery of organic molecules in space was fascinating in showing nature at work on prebiotic chemistry, and although by the early 1990s molecules with up to 13 atoms ($HC_{11}N$) had been found, with the exception of carbonaceous chondrites, the existence of organics even at the level of amino acids had not been proved in outer space. In this sense, most interstellar chemistry, at least as observed by humans, had not reached the level of the Miller–Urey experiment in 1953. And even if reports of detection of the amino acid glycine in molecular clouds at the center of the galaxy turned out to be true, it was a long way from life. With the exception of the organized elements controversy in the early 1960s, and Hoyle and Wickramasinghe's theories (discussed further later on), all of the discussion of organics in outer space environments was strictly prebiotic. Still, even former skeptics were among those claiming organics from interplanetary dust particles and meteorites, and by the end of

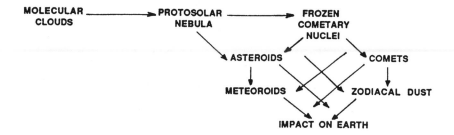

Fig. 6.4. Possible relationships between sources of organic molecules and their delivery to Earth. Today, a horizontal arrow would also be drawn between asteroids and comets, since some asteroids are believed to be defunct comets. From A. Chantal Levasseur-Regourd, "Cometary Studies," in J. Heidmann and M. Klein, eds., *Bioastronomy* (1991), 110. Copyright 1991 Springer Verlag GmbH & Co. KG.

the century there was some consensus that the source of organic material for the beginning of life on Earth could have been outer space. As plausible relationships between extraterrestrial organic molecules and their delivery to Earth were proposed (Fig. 6.4), a refined version of Arrhenius's panspermia gained increasing support.

A much bolder theory of actual life, rather than prebiotic chemicals, coming to Earth was put forth by Fred Hoyle and his colleague Wickramasinghe, who developed several different theories from 1977 to 1981. Rejecting the experimental synthesis of organics since Miller and Urey in 1953 as irrelevant to the conditions present during the origin of life on Earth, in their book *Lifecloud* (1978) the authors argued for the astronomical basis for the origin of life – not just prebiotic molecules, a proposition closer to Arrhenius's original theory but by no means identical. They described how their work on interstellar dust led them to believe that its spectroscopic signature could be explained only by its being composed of cellulose, a polysaccharide biochemical (the substance of wood) that is the most abundant terrestrial organic substance. More specifically, they claimed not only that organic molecules exist, but also that dust clouds form nucleic acids and proteins. These biomolecules were assembled into still more complex forms on planetesimals like comets, which showered the Earth during its first billion years – giving the Earth not only its volatile materials in the atmosphere, but also life in the form of living cells! By 1979 Hoyle and Wickramasinghe had carried their theory a step further, interpreting interstellar absorption features in the ultraviolet spectrum to be evidence of frozen bacterial and viral particles and algae in outer space, microorganisms that occasionally caused epidemics on Earth. By 1981 they were arguing in *Evolution from Space* that the sudden bursts of new life forms on Earth are caused by the arrival of genes from outer space. Their thesis was considered science fiction by most scientists. But the more general concept that events in outer space might well have affected the origin and development of life on Earth was widely accepted, as seen, for example, in the 1985 NASA study entitled *The Evolution of Complex and Higher Organisms*.

Nor was Hoyle's the wildest panspermic theory of the origin of life. Already in 1973, no less a scientist than the Nobelist Francis Crick, of DNA fame, proposed with the biologist Leslie Orgel the idea of "directed panspermia," whereby life was sent by spaceship from other planets to intercept our Earth, among others. Unlike Hoyle and Wickramasinghe, however, Crick and Orgel did not claim that their hypothesis was supported by enough evidence to be taken as true or even likely. It was only a logical possibility, perhaps the ultimate elaboration of panspermic theory that had begun a century earlier.

Philosophical Issues

Aside from the three areas of detection of life on Mars, exobiologists' aspirations for universality in their theories of origin of life, and the discovery of organic molecules in a variety of space environments, the studies of the origin of life and extraterrestrial life were united by philosophical questions underlying all theories of origin – questions that repeatedly thrust themselves to the forefront and loomed even larger in the context of the origin of life beyond the Earth. The old question "What is life?," a property that T. H. Huxley in the 19th century had invested in protoplasm and others in the enzyme, the virus, the gene, or the cell, never ceased to be a concern, even in the terrestrial context. The question of the nature of life was obviously even more critical when applied to extraterrestrial life. Indeed, the nature of life was an important question in designing experiments for the search for life on Mars. Findings toward the end of the century that life on Earth could flourish in deep ocean hydrothermal vents at temperatures of $110°C$, function buried inside rocks, and thrive kilometers below the Earth's surface gave further scope to the possibilities of life beyond the Earth. If "extremophile" bacteria could swim in acid, eat sulfur, and produce methane, and if strange, eyeless worms could live on methane ice deep in the Gulf of Mexico, what might we expect on other planets? By the end of the century, life could no longer be defined in terms of "normal" metabolisms.

The philosophical question that dominated the problem of the origin of life beyond the Earth, however, was not so much the nature of life as the role of chance and necessity in its formation. On the latter question, after all, hinged the whole enterprise of exobiology, for if life was a chance occurrence with very low odds, it need never happen again, no matter how big the universe and how expansive the time available. One might never come to the question of the nature of life in the universe if it could not exist in the first place. In entering that debate, biologists were tackling one of the oldest questions in philosophy; the ancient Greek atomist Democritus had written that "Everything existing in the Universe is the fruit of chance and necessity." From physics to ethics, and occasionally even in early debates on the spontaneous generation of life, this philosophical question loomed but seldom brought consensus.

It is one of the hallmarks of the extraterrestrial life debate that exobiologists, perhaps driven by their desire for a universal biology, placed their faith in the necessity, or at least the high probability, of the origin of life under proper conditions. Eschewing any idea that life was separated from nonlife by a vitalist property, they made the principle of the high probability of the chance formation of life one of their major assumptions, and one they sought to justify as widely held. One sees it already in the first modern treatment of extraterrestrial life by the British Astronomer Royal Sir Harold Spencer Jones, who wrote in 1940 that although the nature of the steps from complex organic substances to the first living cell were unknown, "Nevertheless, it seems reasonable to suppose that whenever in the Universe the proper conditions arise, life must inevitably come into existence. This is the view that is generally accepted by biologists." The same assumption was held by the organic chemist Melvin Calvin, who wrote that all that was needed for an estimate of the probability of cellular life as we know it was an estimate of the number of planets with conditions similar to Earth's. Necessity, or predestination, was the view of many of the exobiological experimentalists working on prebiotic synthesis. And I. S. Shklovskii and Sagan in their volume *Intelligent Life in the Universe* (1966), while admitting that laboratory investigations provided only a "likely story" of the origin of life, nevertheless showed where their sympathies lay by championing the search for intelligent life.

The assumption of necessity adopted by most exobiologists may be traced at least to early in the century in the context of life's origins. Given enough time or space, or a simple enough entity, or the need for only a single first molecule, exobiologists could argue that an event governed by chance was transformed into necessity when the laboratory was the immense, and immensely old, universe. "Wherever life is possible, given time, it should arise," Wald wrote.

But the roots of the opposite view also ran deep. The American Museum of Natural History paleontologist W. D. Matthew had already departed from the view of necessity by 1921, and argued in the context of extraterrestrial life that the chances of life's origins were extremely small. *This* was the position, he held, that was accepted by most biologists. It was the view championed by the French scientist Lecomte du Noüy in his widely read and admired *Human Destiny* (1947). There du Noüy calculated that a single simple protein molecule would take 10^{243} billion years to form by the workings of chance,

243 orders of magnitude more time than was available based on the age of the Earth. That single molecule, he emphasized, was very far from life itself and even more remote from humanity. And in any case, he argued, the laws of chance used in the physical world do not apply to biological phenomena; a materialistic theory of the universe was unsatisfactory. Du Noüy, like others, had his own necessity that denied any role for chance: in examining science and its consequences he concluded that "these consequences lead inevitably to the idea of God." Placed in the broader context of "human destiny," du Noüy's arguments were undoubtedly more influential than all the other technical scientific articles on the subject combined.

The idea of the low probability of the chance formation of life was fortified in the second half of the century as knowledge of the complexity of life increased. Some biologists – even some exobiologists – became increasingly concerned about the odds of life having arisen by chance. In the same paper in which he wrote that "comparing the chemical structure of extraterrestrial life with that of life on Earth" was the only way to answer the question of life's origins, the biochemist Norman Horowitz also wrote that "It is assumed by some biologists, and in my experience, by most astronomers who consider the matter, that the probability of the origin of life given favorable conditions – i.e., conditions resembling those of the primitive Earth – is practically unity. I think that this optimistic estimate may be far from the mark." Because there was as yet no satisfactory theory for the origin of the nucleic acid–protein system found on Earth, Horowitz wrote, "an objective estimate, based on known chemistry and known biology, would lead to a probability for the origin of life of close to zero." Only the discovery of extraterrestrial life might change that estimate, and for Horowitz the Viking results confirmed his skeptical assessment, remarkably similar to that of Matthew 50 years earlier. Those results, however, did not seem to shake the faith of exobiologists. Although a few, like Horowitz, were convinced after the Viking spacecraft failed to find organic molecules on Mars that the conditions for the origin of life were extremely special, most exobiologists put their faith in observation beyond the solar system, preferring not to extrapolate from Mars to the rest of the universe.

The same theme of the workings of chance and its low odds in the origins of life pervaded the influential book *Chance and Necessity* (1971) by the French biologist Jacques Monod. Monod, eventually director of the Pasteur Institute in Paris, and awarded the Nobel Prize with Andre Lwoff and Francois Jacob in 1965 for explaining the genetic regulation of the synthesis of proteins in cells, drew on the latest results in molecular biology to illustrate how chance predominated in all aspects of life's development. In his opinion, life at the molecular level was "the product of an enormous lottery presided over by natural selection, blindly picking the rare winners from among numbers drawn at utter random." Although life was explicable by physical principles (Monod was no vitalist), it was not predictable, either in the fact of its origin or in the fact or direction of its evolution. In Monod's view, the biosphere of the Earth was a unique occurrence that could not be deduced from first principles, no matter how successful a universal theory might be in other domains. If life appeared on the Earth only once, which the unity of biochemistry argued, then the chance of its occurring before the event, its a priori probability, "was virtually zero." Although humanity recoiled from this idea because of a feeling of destiny, Monod warned, that was a feeling against which "we must be constantly on guard." "The universe was not pregnant with life, nor the biosphere with man. Our number came up in the Monte Carlo game," Monod concluded, while not drawing any du Noüyian theological conclusions. Although Monod did not address the subject of life beyond the Earth, if chance reigned on Earth, one could hardly reach any conclusions about life beyond the Earth when any theory of the origin of life could not lead to predictable results.

Similarly, the skeptical biochemist Robert Shapiro, in a chapter on "The Odds" in his *Origins,* calculated the chances of the spontaneous generation of the simplest known organism, the bacterium. Assuming that a billion years was available for the origin of life on Earth and that an ocean 10 km deep covered the planet, he found that 10^{51} trials were available to form the bacterium. Calculating then how many trials would have been required, he concluded that the estimate by Hoyle and Wickramasinghe of $10^{40,000}$ was too low. Rather, he agreed with the Yale University physicist Harold Morowitz that the odds are really 1 in $10^{100,000,000,000}$, or 1 in 10 to the hundred billionth. At these odds, Shapiro remarked, all the time and space in the universe would make no difference; "If we were to wait, we would truly be waiting for a miracle." The only hope was that the origin of life was at a level considerably below that of the bacterium, and this is why the RNA "random replicator" was proposed. However, to create RNA with 20 nucleotides, or 600 atoms rather than the millions required for the bacterium, Shapiro calculated the odds at 1 chance in 10^{992}, still very poor odds indeed. If one accepted these calculations, then the problem of the origin of life on Earth having occurred by the chance assembly of atoms even at the level of the random replicator remained intransigent and the odds of life beyond the Earth extremely low. To those like Hoyle and Shapiro, it seemed that Lecomte du Noüy had *underestimated* just how low the probability was. Still, each could draw his own conclusion. For Shapiro, at least, his calculations did not necessarily mean that the origin of life itself was improbable, but simply that the random replicator mechanism was a poor one; he preferred a more gradual route that did not rely on particular magic molecules.

As some of these same authors had pointed out, if chance and necessity loomed large in the origin of life, its effect was enormously magnified in evolution, which encompassed many steps after life's origin. In his *Time's Arrow and Evolution* (1951), the biologist Harold F. Blum showed how physical nature – in the form of the second law of thermodynamics – placed constraints on the options available for biological evolution. The second law dictated that the universe tends toward disorder, and Blum considered living organisms as thermodynamic systems. The unique place of the Earth in the solar system should make one cautious about life in the universe, he argued. But "if he considers the complexity of living systems and the combination of physical limitations and apparent accident that have characterized the course of organic evolution, he may be still more critical of the idea that life exists elsewhere in the universe." The same argument applied not only to the fact of evolution, but also to the morphology of its products. Even if life did exist elsewhere, "it probably has taken a quite different form. And so life such as we know may be a very unique thing after all, perhaps a species of some inclusive genus, but nevertheless a quite distinct species." Four years later, Blum reaffirmed that belief when he concluded that "For close parallelism of biological evolution among the planets Time's Arrow would have had to play a much more directly deterministic role than now seems likely."

In choosing chance and contingency over necessity and determinism, Blum anticipated the opinion of the majority of evolutionists who would enter the debate in the Space Age. Like Blum, many of them would extend their thinking backward from biological to chemical evolution. The Harvard evolutionist Ernst Mayr, for example, wrote in 1982 that "A full realization of the near impossibility of an origin of life brings home the point [of] how improbable this event was. This is why so many biologists believe that the origin of life was a unique event. The chances that this improbable phenomenon could have occurred several times is exceedingly small, no matter how many millions of planets in the universe." As we shall see in the next section, the issue of chance and necessity would dominate the thinking of Mayr and other evolutionists who commented on extraterrestrial life.

There were, of course, ways around what exobiologists considered such "pessimistic" arguments, and they were championed by more than just exobiologists. In his defense of Darwinism, the Oxford zoologist Richard Dawkins argued that natural selection was like a "blind watchmaker" that, through a cumulative effect, could build complex entities with no purpose in view. And while via cumulative selection the origin of life was not a probable event, and while one could not use its presence on Earth to argue for its probability in the universe, nevertheless, in Dawkins's view, "The origin of life on a planet can be a very improbable event indeed by our everyday standards, or indeed by the standards of the chemistry laboratory, and still be sufficiently probable to have occurred, not just once but many times, all over the universe." The biologist Peter Mora offered a more radical way out when he argued in a paper entitled "The Folly of Probability" that while probability was a proven concept in physics and mathematics, it should not be applied to the question of the origin of life. The latter, Mora argued, might require a teleological explanation taking into account the purpose of living systems. By the 1990s, work on the concepts of self-organization and complexity offered further hope to origin of life optimists. Still, in the view of most exobiologists, the only way to resolve the dilemma of chance and necessity was to search for life beyond the Earth and, after the Viking results, to continue the search.

Clearly, in their search for a universal biology, exobiologists placed their faith in necessity and determinism, even with all of the advances in biochemistry and molecular biology and even in the wake of the Viking results. Although the experimental synthesis of the amino acids and the discovery of complex organic molecules in space environments were clearly not the same as life itself, these syntheses instilled in them a faith that the process would take place.

As we contemplate the evolution of biological complexity as depicted in Figure 6.2, we see how little hard evidence was deduced for the likelihood of extraterrestrial life from the point of view of origin of life studies in the 20th century. Meteors, comets, and interstellar molecules contained only organic compounds at steps 1 and 2. The hope of the early 1960s that meteorites might contain fossil remnants of life at step 4 is still unproven. Artificial synthesis has been achieved for organic molecules at steps 1 and 2 under possible primitive Earth conditions and at the lower levels of step 3 under other conditions. But the assumption of a reducing atmosphere is now in question, and thus the relevance of those experiments to the origin of life. The transition from step 2 to step 3 – the origin of nucleic acids and proteins – is not understood, nor are the transitions from step 3 to step 4 (origin of the cell) or from step 4 to step 5 (origin of intelligence).

Yet, despite the scientific problems and philosophical uncertainties, by the 1990s the four factors that we have documented in this section – the search for life on Mars, the aspiration of exobiologists for a universal biology, the discovery of organic molecules in space, and shared philosophical problems – had consummated the marriage of origin of life and extraterrestrial life studies; their integration was no longer a goal but a fact. Nowhere is this more evident than in a 1990 study on "The Search for Life's Origins" by the National Academy of Sciences, written more than three decades after the same body had expressed concern about planetary contamination – a concern that initiated biologists' interest in life in outer space as a prelude to the search via spacecraft. The "Planetary Biology and Chemical Evolution" aspects emphasized in the report's subtitle were now inseparable from the study of the origin of life. The origin of life study was now dominated by the cosmic history of the biogenic elements and compounds, the implications of early planetary environments for chemical evolution and the origin of life, and even the search for life outside the solar system. Recommendations for further study included spacecraft missions to Mars (still the highest priority despite the Viking results), to comets and asteroids, and to Titan and the outer gas planets; Earth-orbiting facilities for studying the interstellar medium and collecting dust particles; and a raft of related ground-based studies. Experiments in prebiotic synthesis would also continue; the report fully recognized that because of evidence indicating that the primitive Earth's atmosphere may have been carbon dioxide, nitrogen, and water vapor rather than the highly reducing atmosphere assumed by Urey and Miller, "the question of the synthesis of organic compounds on the prebiotic Earth is far from settled and must be reexamined."

These goals – spurred on by the unexpected and still controversial claims in 1996 regarding fossil life in the Mars rock ALH 84001 – would carry the search for life's origins well into the next century. The progress in obtaining data relevant to the origin of life during the 30 years since the dawn of the Space Age, from the study of comets and interstellar matter to the Viking biological experiments, was balanced by the great uncertainties remaining. By the end of the century the quest for the origin of life on Earth was poised to continue, now fully integrated into the goals of exobiology, with a small cadre of biologists still enthusiastically pursuing their quest for universality and for equal status with the physical sciences that a universal biology would bring. In this pursuit, however, they would have to deal with the evolutionary biologists, who were largely pessimistic about the morphology, if not the very existence, of extraterrestrials, placing a damper on the aspirations of those who sought to extend the hard-won principles of terrestrial biology to the universe at large or to incorporate those principles into a more generalized biology.

6.2 EVOLUTION AND EXTRATERRESTRIALS: CHANCE AND NECESSITY REVISITED

If the leap from the synthesis of amino acids to the origin of life was large, the leap from first life to intelligence was, in the eyes of some (though not all), even more monumental. However, it is a matter of terrestrial history not only that chemical evolution begat biological evolution, but also that biological evolution begat intelligence. Beyond the origin of life discussed in the preceding sections, the issue of chance and necessity was at the foundation of three related questions in the evolution of extraterrestrial life: Would intelligences identical to humanity's evolve on other planets? If not, would intelligences different in form evolve? And if the latter, would these intelligences have the ability to communicate with others so different in form? In light of searches for extraterrestrial intelligence (SETI) begun in 1960, these were not merely academic questions, but issues relevant to science policy and to the odds of a return on taxpayers' money. Busy with problems on Earth, evolutionists gave only limited attention to the question of evolution in the extraterrestrial context. Nevertheless, among those who did enter the debate – including some of the pioneers of the evolutionary synthesis – patterns emerged to suggest that the quest for universality in biology had its limits.

Alfred Russel Wallace, cofounder with Darwin of what later became known as the "theory of natural selection," directly addressed the relevance of evolution to extraterrestrials in his book on plurality of worlds at the beginning of the century. In an appendix to the 1904 edition of his book entitled "An Additional Argument Dependent on the Theory of Evolution," Wallace pointed out that the evolution of humanity had depended on millions of distinct modifications. The chances against all these modifications occurring independently, even in two distinct parts of our planet, were "almost infinite." Therefore, on another planet, where environmental conditions are even more diverse, Wallace concluded that the evolution of a species identical to humanity would be "infinitely improbable." Nor, he argued, would any intelligence different in form from humans likely evolve because no other intelligence had evolved on Earth. The evolutionary improbabilities of the development of intelligence on Earth were less than 100 million to 1, he believed, "and the total chances against the evolution of man, or an equivalent moral and intellectual being, in any other planet, through the known laws of evolution, will be represented by a hundred millions of millions to one." To those who agreed that this was true for a universe of natural law, but perhaps not true given the action of a Creator, Wallace argued that we had no knowledge of the Creator's purpose; he might well be satisfied with the millions of souls already produced on Earth and the millions more that would exist in the future.

Wallace felt sure that this argument would "appeal to all biological students of evolution," and in this he seems to have been correct – to a point. Extending his 1921 argument about the unlikelihood of the origin of life on another planet, the paleontologist W. D. Matthew at the American Museum of Natural History in New York agreed with Wallace that the evolution of humanlike creatures on another planet was remote – and constructions like cities or canals even more remote. In the unlikely event that any intelligence did evolve, natural selection guaranteed that "it probably – almost surely – would be so remote in its fundamental character and its external manifestations from our own, that we could not interpret or comprehend the external indications of its existence, nor even probably observe or recognize them." Indeed, beginning in the late 1920s, the many editions of *The Science of Life* by H. G. Wells, G. P. Wells, and Julian Huxley pressed the same point even for nonintelligent life while showing that the search for a universal biology, one conceivable in terms of life on Earth, was not yet underway. Life on Mars would be so different from life on Earth, they argued, that one would have to term it something like Beta Life, "an analogous thing and not the same thing. It may not be individualized; it may not consist of reproductive individuals. It may simply be mobile and metabolic. It is stretching a point to bring these two processes under one identical expression." Moreover, they argued, although one could conceive of silicon instead of carbon or sulfur instead of oxygen acting under different temperatures and pressures to produce consciousness, these would not merely be different forms of life, but would strain the very meaning of the word "life." Nor does this seem to have been a matter of semantics, but rather a statement that the knowledge of life on Earth could tell us little about life in the universe.

Thirty years later, with the unification of biology and the evolutionary synthesis complete, the position of Wallace and Matthew was adopted practically unchanged in terms of the morphology of extraterrestrial life and intelligence but not in terms of the likelihood of its existence. The anthropologist Loren Eiseley was among the first to re-echo the diverse morphology concept, writing in 1957, "Of men elsewhere, and beyond, there will be none forever." Eiseley's conclusion was reached in the context of Darwin, who, he noted, saw clearly that the development of life was not a pattern imposed from without, but one that had been modified by natural selection "along roads which would never be retraced." Expanding to the context of extraterrestrial life, he wrote that "Every creature alive is the product of a unique history. The statistical probability of its precise reduplication on another planet is so small as to be meaningless."

But Eiseley and others from the biological sciences were not unanimous about two other questions: the chances of intelligence occurring in some form and whether one could reach a mutual understanding with such intelligence. The American geneticist H. J. Muller believed that higher lifeforms would develop on other planets, and he agreed that they "may be expected to have followed radically different courses in regard to many of the features" of their evolution, as evidenced by differences among advanced lifeforms on Earth. "How much greater, then, might such differences be between the forms of Earth and those of another planet. These differences, affecting their whole internal economy, including the biochemistry within their cells, would also be expressed in their gross anatomy and in their outer form." Still, Muller believed that such higher lifeforms "would certainly be capable of achieving much mutual understanding with our own, since both had been evolved to deal usefully with a world in which the same physico-chemical and general biological principles operate."

The most detailed and influential treatment of the problem came from the evolutionist George Gaylord Simpson, whose article "The Nonprevalence of

Humanoids," was written as the United States was preparing to spend massive amounts of money to search for life on Mars. Aware of Matthew's 1921 article, Simpson argued that evolutionists needed to be heard by those who espoused exobiology – a " 'science' that has yet to demonstrate that its subject matter exists!" Simpson took to task both physical scientists and biochemists engaged in exobiology who assumed, "usually without even raising the question," that once life arose on another planet its course would be similar to that on Earth. He pointed out that the biologist Harold Blum had distinguished two possibilities: the deterministic one, in which evolution must be similar to Earth's, and the opportunistic one, in which life had many possible courses. Exobiologists, Simpson noted, seemed to fall in the deterministic camp. But he found no support for this conclusion in evolutionary biology on Earth, the only empirical evidence available. To the contrary, he argued that the fossil record showed no necessity that evolution proceed from protozoa to human; rather, most early lifeforms became extinct, indicating that nature experiments with its lifeforms. Moreover, the processes of evolution by mutation and recombination were not directed. "If the causal chain had been different, Homo sapiens would not exist," Simpson concluded, mimicking Wallace's argument 60 years earlier. "No species or any larger group has ever evolved, or can ever evolve, twice." Extending the morphology argument to the more general question of the existence of extraterrestrials, Simpson was pessimistic to the point of ending with a practical suggestion. Pointing to the expenditure for space exobiology he said, "Let us face the fact that this is a gamble at the most adverse odds in history. Then if we want to go on gambling, we will at least recognize that what we are doing resembles a wild spree more than a sober scientific program."

One of the architects of the unification of biology, the evolutionist Theodosius Dobzhansky, while pointing out that chance and necessity were not mutually exclusive in evolution, agreed with Simpson. Writing on the relation of Darwinian evolution to extraterrestrial life in 1972, he pointed out that physical scientists such as Laplace usually adopted the determinist view, but he argued that "Laplacean determinism sheds no light on evolutionary history." Humanity was "invented," not predestined, according to Dobzhansky, and although natural selection would invent many other forms on other planets, the reinvention of humanity was unlikely. A biologist living in the Eocene period, he argued, could not have predicted the emergence of humanity. Of humanity's 100,000 genes, perhaps half had changed since that time, and the chance of the same 50,000 genes being changed in the same ways and in the same sequence was zero. "Natural scientists have been loath, for at least a century, to assume that there is anything radically unique or special about the planet Earth or about the human species. This is an understandable reaction against the traditional view that Earth, and indeed the whole universe, was created specifically for man. The reaction may have gone too far. It is possible that there is, after all, something unique about man and the planet he inhabits."

In the 1980s the views of Simpson and Dobzhansky were carried even further by another pioneer of the evolutionary synthesis, the Harvard biologist Ernst Mayr. Mayr argued not only the unlikelihood of the human form's emerging on another planet, but also "the incredible improbability of genuine intelligence emerging." Even though quantum mechanics had placed determinism in doubt, Mayr was impressed by how physical scientists "still think along deterministic lines," while an evolutionist "is impressed by the incredible improbability of intelligent life ever to have evolved, even on earth." Mayr pointed out that for 3 billion years after the first prokaryotes formed, nothing happened in life. Then when eukaryotes (nucleated cells) originated the Cambrian period, four kingdoms of life developed in quick succession: the protists, fungi, plants, and animals. But intelligence did not even begin to develop in any of the kingdoms except for the animals. Even then, there were hundreds of branching points that led to humanity. These chance events implied not only that humanity was improbable, but intelligence itself: "There were probably more than a billion species of animals on earth, belonging to many millions of separate phyletic lines, all living on this planet earth which is hospitable to intelligence, and yet only a single one of them succeeded in producing intelligence." Nor, he argued, could the "convergent evolution" of an organ such as the eye be applied to intelligence itself because while the history of life on Earth confirms the former, it denies the latter. Finally, Mayr found the idea that extraterrestrials would have the "technology and mode of thinking of late twentieth century man" to be "unbelievably naive." In a remarkable parallel to Simpson's arguments against funding for NASA's life search on Mars, 30 years later Mayr used these evolutionary arguments to argue against NASA's SETI programs.

The Harvard paleontologist Stephen Jay Gould agreed with Wallace and Simpson and other evolutionists about the nonprevalence of humanoids. One of the main points of his book *Wonderful Life* (1989) is that evolution is a "staggeringly improbable series of events, sensible enough in retrospect and subject to rigorous explanation, but utterly unpredictable and quite unrepeatable. . . . Wind back the tape of life to the early days of the Burgess Shale; let it play again from an identical starting point, and the chance becomes vanishingly small that anything like human intelligence would grace the replay." The primary insight won from the Burgess Shale, the richest fossil field of the Cambrian explosion of life about 570 million years ago, is that evolution was a result not of continual proliferation and progress of life, but of the decimation of species "probably accomplished with a strong, perhaps controlling, component of lottery." This view of the contingency of life was bolstered by the increasing acceptance toward the end of the century of the "mass extinction" hypothesis, whereby species were decimated by random occurrences, including the collision of large meteors or asteroids with the Earth.

But at the same time, Gould made the point specifically in regard to extraterrestrial intelligence that lack of detailed repeatability did not necessarily imply lack of extraterrestrials, a distinction some had failed to make. Though some evolutionists (such as Simpson and Mayr) did make that distinction, Gould pointed out that four evolutionists (including himself) had signed a pro-SETI petition in the belief that some finite chance existed of finding such intelligence. He noted that the phenomenon of "convergence" in terrestrial biology, whereby nature had independently invented the general phenomena of the eye, flight, and other adaptive forms many times (though different in detail), might indicate that nature had invented intelligence on many planets. Though some had argued that its appearance only once on Earth indicated the rarity of intelligence as one of nature's general solutions, Gould did not believe this conclusion could be extended from one planet to the universe.

Thus, among evolutionists who cared about the issue (which was by no means most evolutionists), there was a diversity of opinion as to whether intelligence existed beyond the Earth, along with virtual unanimity that if it did exist, the forces of natural selection would produce morphologies vastly different from the humanoid form, as well as uncertainty about the ability of such different intelligent morphologies for mutual recognition and communication. Evolutionists therefore lent credibility to the quest for a universal biology only to the extent that they believed that if life originated, natural selection would operate throughout the universe under whatever conditions life occurred. They did not share the optimism of exobiologists that life would originate, nor take for granted that mutual understanding would be easy or even possible. "We are not alone in the universe, and do not bear alone the whole burden of life and what comes of it," Wald wrote in his 1954 essay. "Life is a cosmic event – so far as we know the most complex state of organization that matter has achieved in our cosmos. It has come many times, in many places – places closed off from us by impenetrable distances, probably never to be crossed even with a signal." In his article on the nonprevalence of humanoids, Simpson made an even more fundamental argument about communication: "I therefore think it extremely unlikely that anything enough like us for real communication of thought exists anywhere in our accessible universe."

On the other hand, exobiologists of the SETI variety, while largely accepting the argument of morphological diversity of the evolutionists, continued their programs despite the uncertainties regarding the existence and communicability of extraterrestrials. Beginning at their earliest conferences, they included discussions in the evolutionary arena on issues like intelligence in dolphins, and went on to broader issues such as interspecies communication and factors affecting the rate of evolution of intelligence. Despite the occasional warning that cognition (as opposed to lower intelligence) could be "exceedingly rare" in the universe, they forged ahead, confident that empiricism was better than speculation. In his Pulitzer Prize-winning book *The Dragons of Eden* (1977), Carl Sagan reflected the view of most exobiologists: "Once life has started in a relatively benign environment and billions of years of evolutionary time are available, the expectation of many of us is that intelligent beings would develop. The evolutionary path, would, of course, be different from that taken on Earth. The precise sequence of events that have taken place here ... have probably not occurred in precisely the same way anywhere else in the entire universe. But there should be many functionally equivalent pathways to a similar end result." It was, Sagan suggested, a matter of natural selection that "smart organisms by and large survive better and leave more offspring than stupid ones," although he admitted that the capacity for self-destruction by technological civilizations left matters uncertain. But the longevity of civilizations aside, exobiologists for the most part accepted contingency in evolution, while still opting for necessity in believing that some form of intelligence would develop.

In their consensus on the contingent nature of the evolution of life, evolutionists and exobiologists alike were in agreement with biochemists who had probed the molecular depths of life, only to find blind chance at work during its origin. Whether in its origin or its evolution, contingency seemed to be the theme of life. While this left wide open the possibilities for extraterrestrial intelligence, unless one accepted the theological conclusion of du Noüy, Hoyle, and Wickramasinghe, the broader implications for terrestrial intelligence were staggering, since the vast majority of terrestrials agreed with du Noüy's general idea of "human destiny." In arguing that life on Earth was contingent, Monod had written, "We would like to think ourselves as necessary, inevitable, ordained from all eternity. All religions, nearly all philosophies, and even a part of science testify to the unwearying, heroic effort of mankind desperately denying its own contingency." Yet this was the one point on which biochemists, evolutionists, and many exobiologists agreed.

Considering the immense difficulties in the study of the origin and evolution of life, and the predilection of evolutionists for natural selection presiding over the blind workings of chance rather than predestination, it may seem remarkable that anyone should seriously consider the possibility of communication with advanced intelligence. Yet the final scientific story remaining to be told in the saga of the extraterrestrial life debate is just this: that the search for extraterrestrial intelligence became a serious scientific project in the last half of the 20th century. There is no greater indication of the strength of the empirical tradition among those exobiologists who supported SETI, of their deeply held basic assumptions about the nature of the universe, and of their continued aspirations for a universal biology, of which terrestrial life was presumed to be only a particular case.

7

SETI

The Search for Extraterrestrial Intelligence

... The presence of interstellar signals is entirely consistent with all we now know, and if signals are present the means of detecting them is now at hand. Few will deny the profound importance, practical and philosophical, which the detection of interstellar communications would have. We therefore feel that a discriminating search for signals deserves a considerable effort. The probability of success is difficult to estimate; but if we never search, the chance of success is zero.

Cocconi and Morrison (1959)

This, then is the paradox: all our logic, all our anti-isocentrism, assures us that we are not unique – that they must *be there. And yet we do not see them.*

David Viewing (1975)

Sing all ye citizens of heaven above!

Adeste Fidelis
Traditional Christmas hymn

All of the problems inherent in the search for planetary systems, in research on the origin of life, and in theories of the evolution of intelligence and technology – in short, all the problems associated with the possibilities of extraterrestrial intelligence – could be leapfrogged if only a method were found for direct communication with such intelligence. It is little wonder, then, that a practical method for doing just that became a kind of holy grail, the focus of a considerable research effort and the source of contentious debate after the dawn of the Space Age. Exactly how this came about is a story full of surprise and drama, an endeavor that eventually opened truly new areas of inquiry in the age-old quest for humanity's place in the universe. That it was challenged by the forces of interstellar colonization, who emphasized the startling observation of the absence of extraterrestrials on Earth to argue that extraterrestrials in fact did not exist, made for an interstellar debate of gargantuan interest for those interested in humanity's status in the cosmos.

Building on knowledge of our own solar system, as well as on the more speculative ideas about other planetary systems and the origin and evolution of life, the search for extraterrestrial intelligence was in many ways a convergence of all previous elements of the extraterrestrial life debate, with the added spectacular dimension of communication and its consequences. It was perhaps foreordained that the search for such intelligence would be a controversial endeavor, and history does not disappoint in this respect. The ultimate in the search for humanity's place in the universe, as it oscillated between imagination and restraint, the search for otherworldly intelligence encompassed all the ambiguities of the extraterrestrial life debate.

7.1 CORNELL, OZMA, AND GREEN BANK: THE OPENING OF THE ELECTROMAGNETIC SPECTRUM FOR SETI

The idea of radio communication on Earth had barely been developed when its pioneers foresaw that the method might be extended even beyond the Earth. Already in 1901 the eccentric American physicist and engineer Nikola Tesla (1856–1943) was convinced, on the basis of his radio experiments, that he "had been the first to hear the greeting of one planet to another." Two decades later, Guglielmo Marconi (1874–1937) generated a headline in the *New York Times,* "Radio to Stars, Marconi's Hope," when he expressed hope for interplanetary communication by radio. In the 1920s, the Amherst astronomer David Todd organized a ground-based project to listen for radio signals from Mars during its close approach to Earth in 1924 (Fig. 7.1). And in 1937, long after most had given up on intelligent Martians, the Harvard astronomer Donald Menzel explored the theoretical possibilities of communicating with the red planet in an article entitled "Can We Signal Mars by Shortwave?"

The idea of interstellar communication – or the Search for Extraterrestrial Intelligence (SETI), as it became widely known after 1976 – had its origins neither in this era of interplanetary communication nor in the space program that had given birth to the Viking search for life on Mars. Rather, interstellar communication found its origin in the "new astronomy," which recognized that celestial objects radiated energy in many regions of the electromagnetic spectrum other than the optical portion familiar to our everyday experience, and that electromagnetic radiation might be useful for interstellar communication.

It was two academic physicists, Giuseppe Cocconi and Philip Morrison at Cornell University, who stumbled onto the subject of interstellar communication almost as an aside to their primary research. Cocconi, a graduate of the University of Milan in 1937, had spent the following year in Rome working with Enrico Fermi building a cloud chamber, one of the early detectors for high-energy physics. Fermi, who received the Nobel Prize that year and escaped to the United States before the outbreak of World War II, was the catalyst inspiring Cocconi to go on in cosmic ray research and particle physics. This he did at Milan and Catania until 1947, when Hans Bethe invited him to join the physics faculty at Cornell. Morrison (Fig. 7.2), on the other hand, was the intellectual son of another theoretical physicist – J. R. Oppenheimer, under whom he received his Ph.D. in theoretical physics from the University of California, Berkeley, in 1940. During the war, Morrison worked at Los

Photo from U. and U.
Army radio operators thruout the country "listened in" for messages from Mars last month—without much success. Corp. John H. Sadler of the Signal Corps is shown at a radio station of the War Department.

Fig. 7.1. The U.S. Army listens for Martian radio signals, according to the plan of David P. Todd, as pictured in *Radio Age* for October 1924.

Fig. 7.2. Philip Morrison, who opened the modern era of SETI with his paper co-authored by Cocconi, seen here in 1975 at about the time of the SETI science workshops that he chaired.

Alamos on the testing and design of the first atomic bomb, an effort headed by Oppenheimer and substantially aided by Fermi. He then joined the physics department at Cornell some 2 years before the arrival of Cocconi. For more than 10 years, Cocconi and Morrison each followed his own research interests until events drew them together in 1959.

Morrison had first pondered the promise of gamma ray astronomy in a 1958 publication. His colleague Cocconi, who had been studying gamma rays emitted by the Cornell synchrotron, one day broached the subject of gamma rays as signals in a conversation with his wife, also a physicist at Cornell. The next day, in the spring of 1959, Cocconi went to see Morrison and asked whether gamma rays could be used for communication between the stars. Morrison suggested that longer wavelengths in the radio region would be better because there would be less interference. Within a few days, Cocconi had calculated that the 250-foot Jodrell Bank Telescope in England – the largest in the world – was just about at the point of being able to signal to or receive Earth-type signals from the nearest star. On the premise that even a radical idea should be called to the attention of the community of scientists, Cocconi and Morrison published what in hindsight turned out to be a landmark paper, "Searching for Interstellar Communications," in the British journal *Nature* in September 1959.

In their paper Cocconi and Morrison admitted that present theories did not yield reliable estimates of the probabilities of planet formation, origin of life, or evolution of advanced societies. But they noted that around one star, the Sun, an advanced society had developed, and that on another (Mars) some type of life might have developed. Despite lack of knowledge of the lifetime of civilizations, the authors found it "unwarranted to deny" that such societies might have very long lifetimes, perhaps comparable to geological time. If so, our Sun would appear as a likely life site, and these societies might be beaming a signal and patiently awaiting an answer. If one accepted this, the problem was to determine on what channel an extraterrestrial civilization might be signaling. Only electromagnetic waves would not be dispersed in the galactic plasma, they argued. But the spectrum of such waves was extremely large. Because of absorption in planetary atmospheres, Cocconi and Morrison quickly narrowed the possibilities to the visible, radio, and gamma ray regions. But because the visible and gamma ray regions would have required too much power or techniques too complicated to be practical, they concluded that the wide radio band from 1 to 10,000 megacycles per second was a rational choice. This was still a broad band to search, but then the authors developed an idea that would become one of the major pivots of the 20th-century SETI debate: "Just in the most favored radio region there lies a unique, objective standard of frequency, which must be known to every observer in the universe: the outstanding radio emission line at 1420 Mc/s (= 21 cm) of neutral hydrogen." The search should begin here, they suggested, observing first some seven Sun-like stars within 15 light years of the Sun.

That the time was ripe for the subject in a broader sense is evident in the fact that Frank Drake, a young astronomer at the National Radio Astronomy Observatory (NRAO) in Green Bank, West Virginia, had independently begun efforts to observe signals such as Cocconi and Morrison described. Here too events conspired to support Drake in what would surely have been seen as a wild scheme at any long-established, conservative institution. This the NRAO decidedly was not. It was the result of a struggle over whether large government laboratories were better than smaller institutions where individuals might have more control over their research. The idea of "large" won out, and in August 1956 the National Science Foundation contracted with Associated Universities Incorporated (which already ran the Brookhaven National Laboratory) to establish NRAO. Groundbreaking took place in October 1957, and in 1958 a new 140-foot telescope was begun.

Into these surroundings Drake came in 1958. A graduate of Cornell and Harvard, Drake had been interested in extraterrestrial life from an early age, and had been influenced during his Cornell years by a lecture on planetary systems given by Otto Struve. Drake was also impressed by an episode from his graduate student days at Harvard in which he detected a strong narrowband signal while observing the Pleiades. Although the signal turned out to originate on Earth, after that Drake always had in the back of his mind the possibility of an artificial signal.

After Drake (Fig. 7.3) came to NRAO, it was natural that he should follow up on this subject. In early 1959 the 85-foot Tatel radio telescope became operational at NRAO while the staff awaited completion of the giant 140-foot one, and in March – just about at the time Cocconi was going to Morrison's office at Cornell – Drake calculated that such a telescope allowed detection of Earthlike radio signals to about 10 light years. It must have been with some trepidation that Drake broached the subject to his colleagues, but this he did one day at the local lunch spot 5 miles from the Green Bank operation. At that lunch was Lloyd Berkner, acting director of NRAO. Berkner supported Drake's project from the beginning: "He was all for it. His whole history had been to push exotic experiments," Drake recalled. "He was a sort of an entrepreneur and gambler of science. He liked to do that kind of thing." Indeed, Berkner would be an early strong supporter of the search for life in the solar system.

By April 1959 the Tatel telescope was completed, but with other tasks, it was summer before Drake and his colleagues began equipment development for their interstellar communication project. At this point another piece of luck occurred: on July 1, Otto Struve, one of the few astronomers of the time who believed intelligent life might be abundant in the universe, became the first director of NRAO. When Drake determined that the Tatel telescope could detect from 10 light years an extraterrestrial signal similar in strength to the Millstone Hill radar system on Earth, Struve was therefore receptive.

In order to avoid criticism, they decided to keep the project as secret as possible and to use mainly equipment useful for more conventional astronomy. In particular, the study of a 21-cm line Zeeman effect, which would indicate the presence of magnetic fields in space, required "two channels, good frequency stability, narrow bandwidths, all very similar to the SETI requirement." The fact that the 21-cm line occurred at an optimum place in the electromagnetic spectrum for detection was, of course, important. Cocconi and Morrison had already pointed this out in their 1959 article, and early the following year, Drake emphasized the same point with a diagram that later

Fig. 7.3. The 85-foot Howard E. Tatel radio telescope used by Frank Drake at the National Radio Astronomy Observatory in Green Bank, West Virginia, for Project Ozma in April 1960. Part of the Ozma team reassembled for the 25th anniversary of the project, including Drake, standing second from the right.

became known as the "microwave window" or "cosmic window." It was a rudimentary form of a diagram that would be seen again and again in this context (Fig. 7.4).

The design of the system – which Drake dubbed "Ozma" because it was searching for distant beings at least as exotic as those in L. Frank Baum's

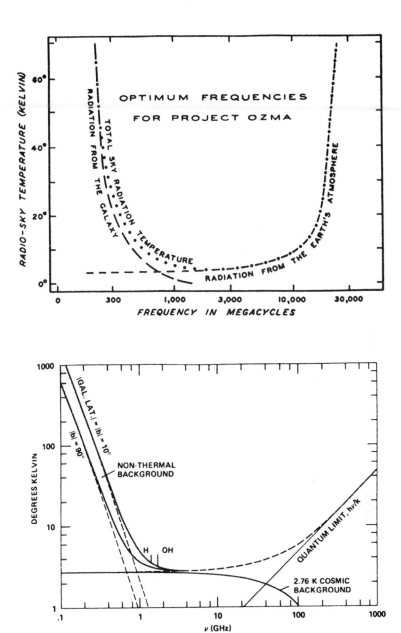

Fig. 7.4. Drake's diagram (top) showing optimum frequencies for interstellar radio communication, bounded on one side by radiation from the Galaxy and on the other by radiation from the Earth's atmosphere. From *Sky and Telescope* (January 1960), p. 141, by permission of Sky Publishing Corp. A later version (bottom) often seen in SETI literature shows more detail but the same "microwave window."

popular *Wizard of Oz* – was relatively simple by present standards. Most important, of course, was the size of the dish – 85 feet. But size alone would not have allowed the program Drake had in mind. His calculation of 10 light years for the detection range depended on the new sensitive radio receivers just coming onto the scene in 1958–1959, among them the maser and the parametric amplifier. Such sensitive receivers also amplified the noise, but the system was designed to reject all signals except for a possible narrowband signal believed to be characteristic of an interstellar signal of intelligent origin. The project was not only low in profile, but also low in budget – only $2000 beyond the conventional parts of the system, including a chart recorder.

Drake and the NRAO staff had been working at a relatively relaxed pace on the equipment for the project when Cocconi and Morrison's paper was published in September. Drake was encouraged because the two physicists had come to the same conclusion he had regarding the strategy of the 21-cm line, on theoretical rather than practical grounds. But Struve was distinctly worried that Green Bank would lose the credit for what he thought was an important idea. Convinced of the need to publicize Ozma, Struve first did so during his Karl T. Compton Lectures at MIT in November 1959 – less than 2 months after Cocconi and Morrison's paper was published. This immediately brought press attention, including Drake's own popular article in *Sky and Telescope* for January 1960 detailing the methods and strategies of Ozma. The new openness also brought more tangible benefits; on reading about the project Dana Atchley, president of Microwave Associates, offered the use of their first copy of a parametric amplifier.

On April 8 Ozma observations began, a routine in which four people participated as observers and several others as telescope operators and technicians. The targets were Tau Ceti and Epsilon Eridani, the two nearest solar-type stars known. On the first day, Tau Ceti was observed until it set at about noon; then the telescope was turned to Epsilon Eridani. "A few minutes went by," Drake recalled. "Then it happened. Wham! Suddenly the chart recorder started banging off scale." The search was not to be that easy; the signal turned out to be due to terrestrial rather than celestial intelligence, and a total of about 200 more hours of observing produced nothing. Although 7200 channels were covered, or 360 kilohertz per star centered around 21 cm in case of Doppler shifts due to the supposed planet's orbital velocity, the remainder of the observations were routine. The project failed to find any intelligent signals of extraterrestrial origin.

Despite its failure, the project fired the imagination of the public, and was widely reported in the press and even in scientific journals. As for the scientific reaction, Struve himself stated in late 1960 that Project Ozma "has aroused more vitriolic criticisms and more laudatory comments than any other recent astronomical venture, and it has divided the astronomers into two camps:

those who are all for it and those who regard it as the worst evil of our generation. There are those who pity us for the publicity we have received and those who accuse us of having invented the project for the sake of publicity." Project Ozma had not reached its hoped-for goal of discovering exotic beings in distant places, but it had served the purpose of stimulating imaginative minds, and an imaginative mind was a difficult thing to stop.

That Struve – nearing the end of his distinguished career – had the courage of his convictions despite the controversy over Project Ozma is evident in his support for an "informal conference" on interstellar communication, held in late 1961 at NRAO. Once again Lloyd Berkner played an important role, for the meeting was sponsored by the Space Science Board of the National Academy of Sciences, a board that Berkner now headed. Moreover, the board included among its members Harold Urey, whose work on origin of life and carbonaceous meteorites had drawn him into the extraterrestrial life debate, and the board's Panel on Exobiology was chaired by Joshua Lederberg. The Green Bank conference may thus be seen as an extension of the Space Science Board's study of the validity of the exploration of life on Mars. Struve served as chairman of the meeting. In addition to the pioneers Cocconi, Morrison, and Drake, the meeting was attended by the astronomers Su-Shu Huang and Carl Sagan (the latter another young member of the Panel on Exobiology), the biochemist Melvin Calvin, the dolphin researcher John C. Lilly, the electrical engineer Bernard M. Oliver, Dana W. Atchley (who had supplied Project Ozma with its parametric amplifier), and J. P. T. Pearman of the National Academy staff. The purpose of the meeting was to examine "the prospects for the existence of other societies in the galaxy with whom communications might be possible; to attempt an estimate of their number; to consider some of the technical problems involved in the establishment of communication; and to examine ways in which our understanding of the problem might be improved." Given the subject matter, it is not surprising that this was an invitation-only meeting that included almost all of the people interested in the subject at the time but excluded the media.

While Pearman handled much of the logistics of the meeting, the organization of the scientific content fell largely to Frank Drake. Thinking in the days before the meeting about how to proceed, he decided to arrange the discussions of extraterrestrial intelligence around an equation that concisely represented the relevant factors. Thus appeared for the first time an equation (eventually known as the Drake Equation) that would be used repeatedly in the following decades in attempts to determine the likelihood of communicating civilizations in our galaxy: $N = R^* f_p n_e f_l f_i f_c L$, where each symbol on the right side of the equation represented a factor on the way to the number of communicating civilizations in the Galaxy (N). The first three factors were astronomical, estimating, respectively, the rate of star formation, the fraction of stars with planets, and the number of planets per star with environments suitable for life. The fourth and fifth factors were biological: the fraction of suitable planets on which life developed and the fraction of those life-bearing planets on which intelligence evolved. The last two factors were social: the fraction of cultures that were communicative over interstellar distances and the lifetime (L) of communicative civilizations. The uncertainties, already great enough for the astronomical parameters, increased as one progressed from the astronomical to the biological to the social.

Although Drake was the first to state these factors in simple equation form, assessments of the probabilities of extraterrestrial life and intelligence had been undertaken sporadically in the course of 20th-century discussions of the subject. On the eve of the events of 1959–1961, former Harvard Observatory Director Harlow Shapley, in his book *Of Stars and Men* (1958), had calculated the number of intelligent civilizations in the universe based on probabilities but had not discussed interstellar communication. Probabilities had also been used by the radio astronomer Ronald Bracewell in another early discussion of the number of advanced communities in the Galaxy. Bracewell, however, had couched his discussion in graphic rather than equation form. And Sebastian von Hoerner had concluded, using probabilities, that 1 in 3 million stars might have a technical civilization but that the longevity of a "technical civilization" (a concept he credited to Bracewell) might be very limited.

When Drake began the Green Bank meeting by writing his equation on the board, he could not have known that he was establishing the paradigm for SETI discussions for at least the last four decades of the 20th century. But by considering in turn astrophysical, biological, and social factors, he did just that, and the Green Bank meeting was only the first of many occasions on which experts would discuss the factors that Drake proposed. In that regard, at the meeting Struve remained enthusiastic about the number of planetary systems, based on his work on stellar rotation, and was supported by his former student Su-Shu Huang, who had concluded from his own research on habitable zones around stars that the number of planets in the Galaxy suitable for life was indeed very large. Calvin, the expert on chemical evolution, argued that the origin of life was a common and even inevitable step in planetary evolution, and his already formidable credentials were given another boost when, during the meeting, he was awarded the Nobel Prize for his work on the chemical pathways of photosynthesis. Lilly, who had just come out with his controversial book *Man and Dolphin,* argued that dolphins formed an intelligent species with a complex language and that we might even be able to communicate with them. Summarizing the results of their discussions, the members of the conference concluded that, depending on the average lifetime for a civilization, the number of communicative civilizations in the Galaxy

Fig. 7.5. Carl Sagan, shown in his earliest days as an Assistant Professor of Astronomy at Harvard, circa 1962.

might range from fewer than 1000 to 1 billion. Opting for optimism, most of the members felt the higher number was likely to be closer to the truth.

By virtue of its conclusions, the Green Bank conference gave the blessing of a group of considerable experts to the theory of Cocconi and Morrison and the observational approach of Drake. Even more than any technical recommendations, it removed to a considerable extent the stigma associated with the idea of intelligence beyond the Earth. "It was wonderful," Sagan recalled, "these good scientists all saying that it wasn't nonsense to think about the subject . . . the fact that they came showed that they didn't think it was beyond the pale. . . . There was such a heady sense in the air that finally we've penetrated the ridicule barrier. . . . It was like a 180 degree flip of this dark secret, this embarrassment. It suddenly became respectable." Energized by the support for this previously "dark secret," Sagan (Fig. 7.5) would go on to become a leader in the broad field of exobiology.

The concept of abundant extraterrestrial intelligence – the rationale that underlay the concepts elaborated and actions undertaken in the previous 2 years – had barely opened the electromagnetic spectrum to SETI, but it clearly opened the door to enormously important questions. As Struve said at the end of the MIT lectures at which he had first announced Project Ozma, "There can be little doubt today that the free will of intelligent beings is not something that exists only on the earth. We must adjust our thinking to this recognition." That was a bold call to action, one of Struve's last before his death in 1963. It was not to go unheeded, even if progress was slow because more sophisticated discussions were needed. Future possibilities seemed endless, but three seminal events – Cocconi and Morrison's paper, Drake's first observations, and the Green Bank conference – laid the foundation for further thought and action.

7.2 A RATIONALE FOR SETI: OPTIMISTS, PESSIMISTS, AND THE DRAKE EQUATION

In the wake of the three seminal events of 1959–1961, two immediate questions faced those who wished to pursue the problem of interstellar communication: (1) What is the likelihood that extraterrestrial intelligence exists? and (2) is interstellar communication via the electromagnetic spectrum really the best method of contact? The Green Bank conference, in the form of the Drake Equation, had given some early tentative answers to the first question and leaned heavily toward the conclusion that the radio search was the optimal method for interstellar communication, at least if civilizations were so numerous that they could be reached with radio telescope technology. The trajectory of the argument within the "radio search" paradigm arced steadily upward until about 1975, when serious doubts were raised about the existence of extraterrestrial intelligence and thus about the propriety of attempts at radio communication. Exactly how the Cocconi–Morrison interstellar communication paradigm became ascendant to 1975, was attacked by the forces of interstellar colonization using some of the radio paradigm's most cherished assumptions, and then remained intact enough to stimulate an increasing number of observational programs is the subject of this section.

The Cocconi–Morrison paper almost immediately stimulated others to elaborate their own ideas. Already late in 1959 the physicist Freeman Dyson proposed an alternative electromagnetic spectrum scenario. On the assumption that other civilizations would be millions of years older than ours, he suggested that they might have redistributed the mass of one of their planets in a spherical shell around their sun for maximum exploitation of their resources. In this case, the resulting object would radiate in the far infrared, a phenomenon amenable to search with existing optical telescopes. This scenario had the advantage of not assuming that advanced civilizations were actively seeking each other, but would passively reveal themselves by their cosmic engineering. Another electromagnetic alternative discussed independently of Green Bank but in the same year was the use of the maser (microwave amplification by stimulated emission of radiation) for interstellar communication. Writing in *Nature*, R. N. Schwartz and the maser inventor, Charles Townes (the latter to receive the Nobel Prize for his work in 1964), concluded that the maser was most advantageously used for such a purpose above the Earth's atmosphere. The following year, however, Bernard Oliver – one of the Green Bank participants – argued that masers were not competitive with radio systems for interstellar communication.

A more radical proposal came from Ronald Bracewell of the Radio Astronomy Institute at Stanford. Unless advanced civilizations were extremely abundant, Bracewell considered it unlikely that we would find any within 100 light years using contemporary technology. Rather, he suggested, they might send probes to the stars. With sophisticated probes, they might explore intensely their neighboring planetary systems and spray the nearest 1000 stars with more modest probes, each going into orbit about a star and equipped with a radio transmitter to attract the attention of any indigent civilization. In this case, Bracewell suggested, we should pay attention only to those civilizations capable of reaching us and search for signs of probes sent to the solar system by these advanced civilizations. Finally, Bracewell suggested that once the probe had found our awakening civilization, we might then be inducted into a "galactic club" of communicating civilizations.

The most radical alternative of all was manned interstellar travel, analyzed in 1962 by Sebastian von Hoerner, a radio astronomer who had been visiting at NRAO during Project Ozma. He concluded that the requirements were so great that "space travel, even in the most distant future, will be confined completely to our own planetary system, and a similar conclusion will hold for any other civilization, no matter how advanced it may be. The only means of communication between different civilizations thus seems to be electromagnetic signals." Edward Purcell, the Nobel Prize winner who, with Ewen, had discovered the 21-cm line, arrived at the same conclusion and ended his article with the memorable words "All this stuff about traveling around the universe in space suits – except for local exploration, which I have not discussed – belongs back where it came from, on the cereal box."

In the Soviet Union a similar pattern emerged favoring the Cocconi–Morrison approach, but with more attention paid to a Dyson-like alternative relying on supercivilizations. In that country the work of Cocconi, Morrison, and Drake had first fallen on the receptive mind of the radio astronomer Iosif S. Shklovskii. Working at the Sternberg Astronomical Institute of Moscow University, in 1960 he expressed in a lengthy article his belief that interstellar communication by radio "is legitimate and timely." This conclusion he based on the collapse of Jeans's hypothesis of the rarity of planetary systems, on Struve's argument that the stellar rotational slowdown at the F5 spectral type was due to the formation of planetary systems, and on the belief that life would develop on planets with suitable conditions. In arriving at his conclusion that "there are in the galaxy at least a billion planets ... on which a highly organized and possibly intelligent life may take place," Shklovskii did not yet make use of the probabilistic arguments that would be characteristic of the Drake Equation. But on the basis of the same qualitative arguments given by Cocconi and Morrison, he did not hesitate to support searches for interstellar communication such as that undertaken by Drake.

Shklovskii's interest soon spread to his colleagues, perhaps in part because the subject of life in space was seen as supported by the Communist philosophy. Shklovskii's book *Universe, Life, Mind,* published in Russian in 1962 on the occasion of the fifth anniversary of the launching of Sputnik, gave prominent attention to the radio search, the first book by an astronomer to do so. The volume stimulated his colleague at Moscow University (and former student), Nicolai Kardashev, to think about the possible levels of advanced societies and, surpassing even the vision of Dyson, to suggest a classification of civilizations into three types. Type I civilizations would have a technological level similar to ours at present, as measured by total energy consumption. Type II civilizations would be capable of harnessing the energy of their own star – for example, the construction of a Dyson sphere. And Type III civilizations would be able to utilize energy on the scale of their own galaxy. Kardashev believed that there was an "extremely low probability" of detecting the Type I civilizations searched for by Project Ozma and suggested that Type II or III civilizations would be better targets. If Shklovskii was the Soviet equivalent to Cocconi–Morrison, then Kardashev was its Drake, but when Kardashev launched his first search in 1963, it was for traces of the more advanced Type II or III civilizations. Although widely influential, Kardashev's classification scheme found no takers in the United States until 1973, when Carl Sagan suggested that it might be better to search for Type II or III civilizations among the nearer galaxies rather than Type I or younger civilizations among the nearer stars.

The appeal to the Soviets of the idea of extraterrestrial intelligence is evident in the conference on Extraterrestrial Civilizations held in 1964 at the Byurakan Observatory in Soviet Armenia. This meeting, which grew out of discussions by Shklovskii's group at Sternberg, including Kardashev, L. M. Gindilis, and V. I. Slysh, differed from the one at Green Bank in that it was attended entirely by radio astronomers. Its purpose also differed in that the aim was "to obtain rational technical and linguistic solutions for the problem of communication with extra-terrestrial civilizations which are much more advanced than the Earth civilization." Accordingly, attention was focused on Kardashev's three types of civilizations, concentrating on the more advanced types while not completely excluding nearby intelligence.

During the 1960s, the Soviets kept up a vigorous debate in SETI. Although they did not discuss alternatives like interstellar travel and probes, as the Americans had, and did not use a quantitative Drake Equation approach, they had much the same discussion in qualitative terms, paid more attention to the likelihood that extraterrestrial civilizations would be much more advanced than ours, emphasized communication rather than search, and gave more attention to the problems of message decoding and problems related to the development of civilizations.

In the United States, by contrast, the discussion after Green Bank centered on the likelihood of communicative extraterrestrial civilizations utilizing radio technology. In this task, the compelling nature of an equation – even one whose parameters were not well known – was not to be denied, as can be seen in the meteoric career of the Drake Equation following its debut at Green Bank. Only a month after the Green Bank meeting in November 1961, Philip Morrison used a similar equation in a NASA lecture. The equation first saw print not in an article by Drake, but in Pearman's account of the Green Bank conference published in a volume entitled *Interstellar Communication* (1963); in the same volume, A. G. W. Cameron used a similar equation. Sagan was also among the first to publish the equation, and Drake himself used it in a paper presented at a JPL symposium on exobiology in February 1963. Although it was not known at first as the Drake Equation, everyone knew (and most stated) where it had first been used. Perhaps the decisive events in the spread of the Drake Equation were Walter Sullivan's popularized account of it in *We Are Not Alone* (1964) and Sagan's incorporation of it into his translation and expansion of Shklovskii's book *Intelligent Life in the Universe* (1966), which became the bible of the SETI movement. These books assured the rapid diffusion of the Drake Equation to public and scientist alike.

Although not immediately used in the Soviet Union, the Drake Equation, with its emphasis on radio communication, focused attention on the electromagnetic radio search paradigm. Already by 1966 this concept, and all of the assumptions that went with it, were sufficiently entrenched that Dyson labeled it the "orthodox view" of interstellar communication, characterized not by interstellar travel, but by "a slow and benign exchange of messages, a contact carrying only information and wisdom around the galaxy, not conflict and turmoil." As anyone who read science fiction knew, this was not the only possible view of the universe. But it was a practical method, a logical extension of the new field of radio astronomy, and one that at least some of its practitioners were keen to carry out. For these reasons, the discussion of rationale and strategy within the radio search paradigm continued its climb.

Nowhere is this more evident than in the first international SETI meeting, at which Sagan also played an important organizing role. Held 10 years after the Green Bank meeting, its organizing principle was the Drake Equation, which for all its quantitative look showed how difficult still were concrete results. Held at the Byurakan Astrophysical Observatory in Yerevan, USSR (in sight of Mt. Ararat), the meeting was sponsored by the Academies of Sciences of both the United States and the Soviet Union. The organizers included not only Sagan, Drake, and Morrison in the United States, but also Ambartsumian, Kardashev, Shklovskii, and Troitskii of the Soviet Union. Instead of the 11 participants at Green Bank in 1961, 28 Soviets, 15 Americans, and 4 from other nations participated – and from a wide variety of disciplines. The conclusion of the group – very tentatively expressed in *Communication with Extraterrestrial Intelligence* – was that perhaps a million technical civilizations existed in the Galaxy. Although the participants also discussed astroengineering activity, the further discussions of techniques of contact, message content, the consequences of contact, and the recommendations focused on the radio search method.

While the Byurakan meeting was in progress, in the United States NASA's interest in extraterrestrial intelligence was slowly beginning to stir – with what effect we shall see in the next section. The fruits of their first study, headed by the Green Bank veteran Bernard Oliver and the NASA scientist John Billingham, became known as Project Cyclops, whose rationale also relied heavily on the Drake Equation. Oliver, employing what he frankly called some "very approximate and probably optimistic values" in the equation, came to the conclusion that the number of communicative civilizations was approximately equal to the average lifetime of galactic civilizations – and the latter was, of course, wide open to conjecture. As to the question of where to search, the Cyclops report, while admitting that "the arguments for using the hydrogen line no longer seem quite so compelling" now that many other radio frequency lines had been discovered, nevertheless confirmed the concept of a "naturally identified frequency" as a method for narrowing the search. It argued that the region of spectrum from 1 to 3 gigahertz (GHz, 1 to 3 billion cycles per second) was the location of "likely beacon frequencies," in particular the portion from 1.420 GHz (the 21-cm hydrogen line) to 1.662 GHz (the lowest of the hydroxyl lines). While leaving open other possibilities, Oliver argued that "surely the band lying between the resonances of the disassociation products of water is ideally situated and an uncannily poetic place for water-based life to seek its kind. Where shall we meet? At the water hole, of course!" Combined with the strategy of "magic frequencies," and especially the "water hole," optimistic values in the Drake Equation became a common rationale of most discussions of the subject during the first decade of its career (Table 7.1). As we shall see, and as the last entry in Table 7.1 indicates, there was another side to the story amply represented a decade later.

By 1971, then, the Drake Equation had become a common feature of the SETI movement, and had succeeded in persuading not only the public but

Table 7.1. *Estimates of factors in the Drake Equation for communicative civilizations*

Author	Date	R^*	f_p	n_e	f_l	f_i	f_c	L	N
Green Bank	1963	1–10	.5	1–5	1	1	.1	10^3–10^8	$< 10^3$–10^9
Cameron	1963		1	.3	1	1	.5	10^6	2×10^6
Sagan	1963	10	1	1	1	.1	.1	10^7	10^6
Shklovskii/ Sagan	1966	10	1	1	1	.1	.1	10^7	10^6
Byurakan	1971	10	1	1	—————.01—————			10^7	10^6
Oliver	1971	20	.5	1	.2	1	.5	?	$= L$
Rood/ Trefil	1981	.05	.1	.05	.01	.5	.5	10^4	.003

R^* = rate of star formation, f_p = fraction of stars forming planets, n_e = number of planets per star with environments suitable for life, f_l = fraction of suitable planets on which life develops, f_i = fraction of life-bearing planets on which intelligence evolves, f_c = fraction of intelligent cultures communicative over interstellar distances, L = lifetime of a communicative civilization, N = number of communicative civilizations in the Galaxy at a given time.

Note: For references see Steven J. Dick, *The Biological Universe*, p. 441.

also many astronomers of the likelihood of extraterrestrial intelligence. Although there were counterproposals to the radio method, one measure of the seriousness with which some in the scientific community took the existence of extraterrestrials by the mid-1970s is evident in the reaction of the Nobel Prize-winning astronomer Sir Martin Ryle. When in 1974 Drake and other astronomers at Arecibo sent a message to some 300,000 stars that comprise the Great Cluster of Hercules (also known as M 13), Ryle objected strenuously. Although the targets were some 25,000 light years distant, Ryle was acting on a deeply felt principle when he agitated for the International Astronomical Union to urge that no attempts be made to communicate with other civilizations because of possible hostile consequences.

Although suggestions were made for the modification of the Drake Equation, it continued to be used in substantially unchanged form. Users realized its weaknesses, and critics were quick to point them out. The equation, Bracewell noted, boiled down to two factors – the rate of formation of intelligent communities and their longevity, about neither one of which anything was known. Even Oliver referred to it as "a way of compressing a large amount of ignorance into small space."

However shaky its status as a scientific tool, as Shklovskii later said, it was not devoid of meaning. The goal was to turn its subjective probabilities into mathematical probabilities, and one could see this beginning to happen in the case of at least one of the factors: the number of planetary systems. But most of the other parameters were admittedly beyond the realm of 20th-century science. Faced with this dilemma, rather than withdraw from the debate, many preferred to fall back on the probabilistic arguments so prevalent in the origin of life debate. Green Bank Director Otto Struve, for example, held that "an intrinsically improbable single event may become highly probable if the number of events is very great." For Struve "this conclusion is of great philosophical interest. I believe that science has reached the point where it is necessary to take into account the action of intelligent beings, in addition to the action of the classical laws of physics." Just this problem – taking into account the action of intelligent beings – was to become the central focus of the next decade of debate, with unexpected results for SETI.

Although the persistence of optimists and pessimists from the beginning of the debate is evident in the varying values and opinions about the Drake Equation, after 1975 SETI went through what one of its participants called "a major crisis of identity and purpose." This crisis undermined the very foundations of the SETI endeavor, bringing into question the logic of the radio search paradigm by claiming that all searches of the electromagnetic spectrum might well be fruitless.

The origin of the crisis was renewed attention to a question casually raised by the pioneering nuclear physicist Enrico Fermi almost 10 years before the modern era of SETI. During lunch with colleagues at Los Alamos in 1950, Fermi had simply asked, "If there are extraterrestrials, where are they?", a question that now became known as the Fermi Paradox. UFO believers would have answered without hesitation that the extraterrestrials *were* here – they had known it all along – and perhaps it was in part the reputation of this group that kept scientists from pursuing the Fermi question for 25 years. That in 1975 the issue was raised in forceful form in two independent articles in the United States and Britain, without at least one of the authors knowing about Fermi's casual question, is some indication of the force of its logic. That there are no intelligent beings from outer space on Earth now, argued Michael Hart and David Viewing in their respective articles, is an observational fact that argues strongly that extraterrestrials do not exist. The basis for this conclusion was the assertion that interstellar travel was possible after all, coupled with attention to the time scales involved. The pessimistic views of those like Purcell, Hart claimed, are based on relativistic spaceflight; the use of nuclear propulsion at, say, 1/10th the speed of light would have much more reasonable energy requirements. Given the age of the universe and the time needed for intelligence to develop, Hart and Viewing proposed, extraterrestrials should have populated the galaxy. At a velocity of 1/10th the speed of light, Hart argued, this would have occurred in a mere 1 million years. This applied even if only one race existed in the Galaxy. We do not see them here; therefore, they do not exist, and "an extensive search for radio messages from other civilizations is probably a waste of time and money."

The prospect of interstellar colonization rather than interstellar communication was stunning, and others soon joined the looming battle of ideas, some on the optimistic side (from the SETI viewpoint), others on the pessimistic side. By 1979 the battle lines were drawn, and the interstellar colonization forces felt confident enough to hold a "Where Are They?" conference centered on the Fermi Paradox. Most of the participants undoubtedly agreed with the conclusion that extraterrestrials were not on Earth. But if extraterrestrial civilizations were abundant, why were they not on Earth? The presence of several of the SETI pioneers testified that the question was a serious one. The radio astronomer Bracewell reiterated his belief that probes would certainly have covered the Galaxy by now. The physicist Dyson reviewed the propulsion systems that might be available to advanced civilizations. Perhaps most impressively, Sebastian von Hoerner, who in 1962 had concluded that space travel would be confined to the solar system of any given civilization, now found Hart's question a "great puzzle . . . still unsolved." Newer participants also added their puzzlement, especially in light of the contemporary debate about Gerard K. O'Neill's space colonies. The physicist Eric Jones, having run computer simulations of the expansion of a spacefaring civilization in the Galaxy, concluded that on any reasonable assumptions extraterrestrials should be here. That they are not, he suggested, gave credence to the claims of Hart and Viewing that no other civilization had arisen in the Galaxy. Other alternatives were possible, including Michael Papagiannis's suggestion that extraterrestrials might be in the asteroid belt or that the evolution to intelligence takes much longer than is usually accepted. But the conference raised the very real possibility that we were alone in the Galaxy. In what amounted to a healthy dose of skepticism, the "Where Are They?" participants abandoned the Drake Equation and what it had come to stand for – the abundance of extraterrestrial civilizations.

None took the "Hart attack" as seriously as mathematical physicist Frank Tipler, who in a barrage of articles beginning in 1981 took the extreme position that the arguments were so compelling that it was a waste of taxpayers' money to undertake a search. Any civilization that had developed the technology for interstellar communication, he argued, must also have developed the technology for interstellar travel. Moreover, any species capable of interstellar communication would also be adept at computer technology and would have developed "a self-replicating universal constructor with intelligence comparable to the human level," something some experts on Earth expected within a century. Such a machine would have explored or colonized the Galaxy within 300 million years, he argued, at a cost less than that of operating a microwave beacon for several hundred years, as SETI advocates postulated alien civilizations might do.

Although some had immediately rejected the ideas of Hart, Viewing, and Tipler, most saw their force, including some in the SETI community. In 1979, 10 weeks before the "Where Are They?" conference, the SETI community addressed Hart's thesis in an open forum at the triennial meeting of the International Astronomical Union in Montreal, a meeting attended by Shklovskii, who himself seemed convinced by the Fermi Paradox. While agreeing with Hart that the Galaxy would be colonized in 10 million years were interstellar colonization feasible, Drake argued that interstellar travel and colonization are too expensive; in short, "the workings of biology, the physical laws of energy and the vast interstellar distances conspire to make interstellar colonization unthinkable for all time." The extraterrestrials, he added, "are living comfortably and well in the environs of their own star." By 1984 the furor raised by the galactic colonization thesis reached its peak at another International Astronomical Union meeting at which the previously expressed opinions were consolidated. Table 7.2 summarizes the results of two decades of explanations for the apparent absence of interstellar colonization.

The interstellar colonization controversy thus had a sobering, but not fatal, effect on the concept of interstellar communication. By arguing that galactic colonization should already have taken place, it forced greater attention to be given to search alternatives, ranging from Bracewell or Tipler probes to Dyson spheres and Kardashev's Type II or III civilizations. SETI proponents were perhaps more receptive to alterations in the standard radio paradigm, including the proposal in 1979 of the University of Washington astronomer Woodruff T. Sullivan III and his colleagues that eavesdropping for radio leakage should be considered in addition to the search for purposeful signals. But most of all, SETI advocates did not abandon their conviction that the search should continue, an affirmation of the primacy of observation over theory in science. In the Soviet Union, however, without the help of Shklovskii, observational programs suffered severely.

If the optimists were to pursue the observational goals of SETI, they would have to deal with the problems of equipment, personnel, and money – resources that might be used for much surer bets than interstellar communication. In short, whether seeking support from a university or the federal government, on a small or large scale SETI would have to enter the arena of science policy. To the practical considerations of that arena, from the airy heights of rationale, we now turn.

Table 7.2. *Explanations for the apparent absence of extraterrestrials on Earth*

Extraterrestrials exist but								Extraterrestrials do not exist
No interstellar travel because	Travel is slow	Interactions slow colonization rate	They are undetected	Limits to growth	Too far	Lack of interest/ persistence		

No interstellar travel because:
- Too expensive: Drake (1980)
- Physically impossible: Drake (1985), Wolfe (1985)
- Too hazardous: Wolfe (1985)

Travel is slow: Newman & Sagan (1981)

Interactions slow colonization rate: Turner (1985)

They are undetected: Ball (1973), Papagiannis (1978), Stephenson (1979)

Limits to growth: Morrison (1985)

Too far: Wesson (1990)

Lack of interest/persistence: Finney (1985), Wolfe (1985)

Extraterrestrials do not exist: Hart (1975), Viewing (1975), Hart (1980), Tipler (1980), Jones (1985), Hart (1985), Tipler (1986)

Note: For references see Steven J. Dick, *The Biological Universe*, p. 451.

7.3 A STRATEGY FOR SETI: THE DEVELOPMENT
OF OBSERVATIONAL PROGRAMS

The widespread acceptance of the argument for the primacy of observation over theory in science is evident in the fact that while some 15 observational searches occurred in the 15 years prior to 1975, more than 40 were undertaken on various scales in the following 15 years – even after serious theoretical doubts had been raised about the existence of extraterrestrials. In contrast to discussions of the rationale for SETI, actual searches required a commitment to specific strategies – of hardware and software, of frequencies, apertures and bandwidths, of financial commitments and personnel – in short, all the considerations Drake had to contend with more or less unilaterally during the days of Project Ozma in 1960, now multiplied many times over in cost and complexity. While Soviet SETI observing programs were severely affected by the interstellar colonization crisis, American programs flourished, often inspired or financially aided by NASA. But ironically, in direct proportion to the influx of government funding, SETI became vulnerable to the always unpredictable and sometimes irrational political winds of the U.S. Congress, ultimately with disastrous results for NASA's own flagship SETI program.

Perhaps significantly, the majority of searches prior to 1975 were undertaken in the Soviet Union – an indication of the influence of the discussions inspired by Shklovskii and his radio astronomy group. As we have seen in the last section, while the United States concentrated on the more immediate problems of its space program, including life in the solar system, discussions on interstellar communication were undertaken in the Soviet Union, first at the all-union conference at Byurakan in 1964 and then (with American help) at the international meeting at the same location in 1971. Aside from a search by an American astronomer observing one galaxy from Australia, the Russians had a monopoly on searches for interstellar communication until several were carried out at NRAO in Green Bank in the early 1970s.

To judge by their observations, the Soviets were not overly impressed with the strategy of the 21-cm line. Impressed instead with the premise that other civilizations were likely to be millions or billions of years more advanced than us, they turned their attention to the effects such civilizations might produce. Thus, when in 1963 Kardashev and Sholomitskii carried out the first search after Ozma in the wake of Kardashev's paper on Type I, II, and III civilizations, it was at 920 megahertz (MHz, or millions of cycles per second) rather than the 1420-Mhz equivalent to the 21-cm line. For a few brief weeks, the strategy seemed to have paid off with a spectacular discovery that soon became a part of SETI legend. Concentrating on peculiar radio sources of very small angular dimensions, Kardashev and Sholomitskii claimed that they had actually detected a possible artificial signal from a Type III civilization associated with a variable radio source designated CTA-102. Although the incident caused a great uproar, it turned out to be a quasar with a large red shift. Subsequent Soviet searches in the late 1960s and early 1970s also largely shunned the 21-cm strategy.

The first searches in the United States had to wait 10 years after Ozma. When they did resume, they clearly followed in the footsteps of Drake and were also undertaken at the NRAO in Green Bank. Both programs there in the early 1970s observed selected stars using the 21-cm hydrogen line and the 140-foot or 300-foot Green Bank telescopes compared to Drake's 85-foot one. A third American search, which has turned out to be the longest-running of all, began under John Kraus and Robert S. Dixon at Ohio State University in 1973 and again used the 21-cm strategy. Utilizing a meridian-transit telescope with a collecting area of 2200 square meters (equivalent to a parabolic dish 175 feet in diameter), the observing program made use largely of student volunteers. Over the course of its program, the Ohio State telescope detected a number of interesting transient signals, most notably one observed in 1977 known as the "WOW signal" after the exclamation penned on the observing record. But none lasted long enough for positive identification. Ironically, Drake himself, along with Carl Sagan, was the first in the United States to stray from the single-star 21-cm strategy, if only temporarily. Using the 1000-foot dish at Arecibo, in 1975–1976 they observed four galaxies while searching for Kardashev's Type II civilizations.

The year 1975 – during which the idea of interstellar colonization began its career in earnest – also marks a shift in observing programs; this was the year the United States began its domination of SETI observing programs, while the Russians almost completely dropped out. Although the Russians had developed a systematic observing program in 1974, and later proposed to set up an array of 100 1-meter dishes for SETI, neither materialized, at least in part because of the change in heart of Shklovskii regarding the likelihood of life. At the 1991 U.S.–USSR conference on SETI, not a single SETI radio observing program was reported from the Russian side.

In the United States, on the other hand, Ohio State continued its program; new projects were initiated at the NRAO, Arecibo, and Harvard observatories; and NASA began planning for what would become the flagship SETI program. Notably, the University of California, Berkeley-based Project SERENDIP (an acronym for Search for Extraterrestrial Emission from Nearby Developed Intelligent Populations) was initiated in the late 1970s in a "parasitic" mode, that is, as a "piggyback" program operated in tandem with other regularly scheduled observing programs. At Arecibo, where Drake served as director, a number of small programs were undertaken beginning in 1977. Among these was a standard 21-cm approach of Paul Horowitz, which grew into Projects Sentinel and META (Megachannel Extraterrestrial Assay) at

Harvard. Funded in part by the Planetary Society, the largest space-advocacy group in the world (headed by Sagan), the program was expanded to META II in Argentina in 1990. Horowitz's programs held the record for frequency resolution of .015 Hz.

By 1985, a quarter-century after Drake's first search, NASA's future SETI Project scientist Jill Tarter could distinguish three SETI strategies in terms of telescope usage: the usual directed searches in which telescopes could be used for brief SETI observations; shared searches such as SERENDIP, which operated in a parasitic mode during other observations or analyzed old signals; and dedicated searches such as the Ohio State effort, in which an instrument was devoted exclusively to SETI over an extended period. These observations had been undertaken with radio telescopes ranging from a few tens of meters in diameter to the giant 305-meter instrument at Arecibo, at a variety of "magic" frequencies based largely on fundamental line radiations from atoms or molecules, with frequency resolutions varying from megahertz to the few hundredths of a hertz of Horowitz's Sentinel Project at Harvard. Most had taken only a few hours or tens of hours; of the 120,000 SETI hours logged by 1985, 100,000 had occurred at the two dedicated facilities at Ohio State and Harvard. As the characteristics of selected observing programs summarized in Table 7.3 show, only a few observations had been undertaken at optical or infrared frequencies; the vast majority were in the radio region originally favored by Cocconi, Morrison, and Drake. Most impressively, Tarter's analysis demonstrated the accelerating pace of observations; 25 of the 45 entries in her summary of observing programs were searches conducted since 1979, and seven countries were involved instead of three. By the 1990s, SETI had become a global if still sporadic effort.

Of all the search programs, NASA's was to be the most comprehensive, and the development of its program illustrates on a large scale many of the problems that all SETI programs faced. There was, first of all, the problem of marshaling resources and funding, and here two figures played a key role. John Billingham, a physician at NASA's Ames Research Center, convinced Ames management in 1970 that interstellar communication was within the scope of NASA's mission to the extent that it sponsored a mini-study. The optimistic results of that study led in 1971 to a more ambitious "design study of a system for detecting extraterrestrial intelligent life." Known as Project Cyclops, the study was undertaken not as a major NASA project, but as part of a summer faculty fellowship program in engineering systems design sponsored by Stanford, NASA, and Ames.

The key figure in the Cyclops study, and the author of its influential report, was Bernard M. Oliver, an electrical engineer, vice president of Hewlett-Packard, and member of the 1961 Green Bank "Order of the Dolphins," whom we encountered earlier in this chapter. Oliver and his colleagues envisioned

Table 7.3. *Characteristics of selected SETI observing programs*

Date	Observer	Site	Instrument size (meters)	Search freq. (MHz)	Frequency resolution (Hertz)	Objects	Total hours	Comments
1960	Drake	NRAO Green Bank	26	1420–1420.4	100	Two stars	400	Project Ozma, 1 channel
1963	Kardashev/ Sholomitskii	Crimea DSS	16 (8 antennas)	920	10 MHz	Two quasars	80	CTA 102 – Type III civilization reported
1969–1983	Troitskii/ Bondar/ Starodubtsev & others	Gorky Crimea	Dipole	1863, 927, 600		All-sky	1200/ year	Search for sporadic pulses
1973–1997	Dixon et al.	Ohio State	53	1420.4	10KHz 1 KHz	All-sky		Longest-running SETI program
1978–	Shvartsman et al.	Zelen-chukskaya	6	Optical		30 radio objects		Optical search for short pulses
1975–1976	Drake/ Sagan	Arecibo	305	1420, 1667, 2380	1000	Four galaxies	100	Search for Type II civilizations in galaxies
1976–	Bowyer et al.	Hat Creek and others	26, etc.	1410ff. 1653ff.	2500	All-sky	C^a	SERENDIP piggyback system
1983–	Horowitz et al.	Harvard	26	1420.4, etc.	.03	All-sky	C	Sentinel – "suitcase SETI" followed by META (1985)
1990–	Colomb et al.	Argentina	30	1420.4	.05	All-sky	C	Southern Hemisphere META II
1990–	Betz	Mount Wilson	1.65	10 microns (infrared)	3.5 MHz	100 stars	C	IR search interferometer
1992	NASA	Arecibo	305	1–3 GHz	1–28	Targeted	< 1 yr	HRMS/MCSA[b] terminated by Congress
1992–1993	NASA	JPL/DSN	34	1–10 GHz	30	All-sky	< 1 yr	HRMS/WBSA[c] terminated by Congress
1995	SETI Institute	Parkes, Australia	64	1–3 GHz	1–28	Targeted		Phoenix: HRMS descendant
1996–	SETI Institute	NRAO Green Bank	47	1–3 GHz	1–28	Targeted	C	Phoenix: HRMS descendant

[a] C = continuing program.
[b] High Resolution Microwave Survey/Multichannel Spectrum Analyzer.
[c] High Resolution Microwave Survey/Wide Band Spectrum Analyzer.
Source: Adapted from Jill Tarter, by permission of the author and Reidel Publishing Company.

Fig. 7.6. Artist's conception of a portion of the Cyclops array, proposed as a result of the NASA/ASEE summer study in 1971, showing antennae and the central control and processing building at the right. Courtesy NASA Ames Research Center.

an "orchard" of perhaps 1000 100-meter antennas covering a total area some 10 km in diameter (Fig. 7.6). But considered as a "phased array" of connected antennas, any construction could start out small and sequentially add more antennas in the event that no signals were detected. The study addressed details of such aspects of the project as antenna elements, receiver systems, and signal processing, as well as more general problems about the probability of life in the universe and search strategies. The resulting publication, which emphasized the "water hole" frequency as the most likely place to search, holds an interesting place in SETI history: although the recommendations for the full project (requiring some $6–10 billion over 10–15 years) were much too ambitious for NASA funding, the study not only served as a demonstration of what could be done, but also demonstrated the technical feasibility of interstellar communication and provided a benchmark against which smaller programs could be measured.

Ironically it was just as the interstellar colonization crisis of SETI was beginning that NASA began its first major study of a realistic SETI program. Most important were a series of six Workshops on Interstellar Communication held in 1975–1976 under the chairmanship of Philip Morrison. These workshops, and three offshoots dedicated to planet detection and the evolution of technological civilizations, proved collectively to be a landmark in SETI history and were critical in stimulating interest and support in the wider scientific community. Having considered interstellar travel, robot probes, and electromagnetic signals, the workshops confirmed that radio signals were the optimum method for communication. But they also recognized that the extremely large number of frequencies – analogous to a very extended radio dial – required the search to be limited in direction or frequency or both. Although no consensus was reached on search strategy, the report of the workshops provided the first public discussion of a possible "bimodal" method for the search, to include both a targeted search (advocated by Ames) and an all-sky survey.

The assumptions behind the targeted and all-sky strategies would be the subject of much further discussion. In fact, the departure from the Ames targeted search strategy was the result of a vigorous campaign by JPL Director Bruce Murray. JPL, an agency funded by NASA through Caltech, ran the Deep Space Network and had the expertise in radio astronomy needed for SETI. By 1977 JPL too had a SETI office, and what emerged was an Ames–JPL partnership that would become a major feature of NASA's formal SETI program.

With the impetus of the Morrison workshops, NASA's attention turned to an actual program that might be funded. SETI would be a program significantly unlike most NASA endeavors. No spacecraft would be built, no launch risks encountered, no possibility of equipment failure in space. SETI was to be a ground-based program, and political and economic realities dictated that it would be no Cyclops with a vast array of new equipment; the embryonic program would use existing radio telescopes to which would be attached specialized detectors, and it was the detectors that would be the main object of funding. The proposed total cost of the SETI program, including 5 years of research and development (R&D) and 10 years of the operational phase, would be about $100 million, 10 percent of the billion-dollar Viking project but roughly equal to the cost of Viking's biological experiments.

With the possibility of significant NASA funding on the horizon, in June 1979 NASA sponsored a major conference at Ames on "Life in the Universe." With the impetus provided by the Morrison workshops, NASA by this time had formally adopted a search strategy – the bimodal strategy that not only made sense scientifically, but also satisfied the desire of both JPL and Ames to work on the project. Scientists at both Ames and JPL therefore authored the paper given at the 1979 conference, the first to lay out the NASA program in detail. Terming it "a modest but wide ranging exploratory program," the authors described a 10-year effort "using existing radio telescopes and advanced electronic systems with the objective of trying to detect the presence of just one signal generated by another intelligent species, if such exists." JPL would undertake the all-sky survey, at wavelengths ranging from 1.2 to 10 GHz, while Ames would concentrate with more sensitivity on the targeted search among some 700 stars within 25 parsecs. This effort, according to the group, was made possible by a maturing radio technology, recent digital solid-state advances for the detectors, and "a minimum number of ad hoc assumptions." The authors contemplated a 10 millionfold increase in capability over the sum of all previous searches, and the equipment to accomplish this would be the focus of the R&D effort. Known as the Multi-Channel Spectrum Analyzer (MCSA), this detector – along with its software algorithms – was the heart of the system, the means by which the "cosmic haystack" could be searched for its hidden needle, a favorite metaphor employed by NASA. A graphic representation of the cosmic haystack in this article first dramatically depicted the magnitude of the task (Fig. 7.7). By 1979, then, NASA had a detailed idea for a coherent SETI program but not much money to carry it out.

As a result of all this discussion, beginning in the early 1980s NASA's Ames and JPL facilities were finally able to embark on an intensive program, known initially as MOP (Microwave Observing Program) and after 1992 as HRMS (High Resolution Microwave Survey), to build the instrumentation necessary for a systematic search for intelligent life. A 5-year R&D phase at the level of about $1.5 million per year was carried out from 1983 to 1987. Following a period of uncertainty and minimal funding, in 1990 SETI took on the status of an approved NASA project, no longer in the R&D phase, and entered a 10-year phase of final development and operations, to be completed by the new millennium at a total cost of about $108 million.

The final program that emerged in 1992 was quite similar to that envisioned in 1979. The Ames Targeted Search Element of the NASA SETI program was to search for 800–1000 solar-type stars within about 100 light years. Beginning with Arecibo, it would use the largest radio telescopes possible, observe each star for 300–1000 seconds, and focus on the 2 billion Hz in the 1- to 3-GHz region of the microwave spectrum. Because of practical limitations, each star had to be observed 100 times to cover the entire 2 GHz. JPL's Sky Survey Element, on the other hand, made no assumptions about specific preferred targets in the sky, but observed the entire sky at 1–10 GHz with smaller, 34-meter class, radio telescopes beginning with those of the Deep Space Network. The Sky Survey's observational strategy was to examine each spot in a tessellated "racetrack" pattern for only a few seconds at most, decreasing sensitivity by several orders of magnitude and losing the ability to detect any pulsed transmissions over time periods longer than its observation at a single spot. Each mosaic would build up a "sky frame," and approximately 25,000 sky frames would be required to cover all directions and frequencies, each taking about 2 hours to complete, for a total of about 7 years for the complete

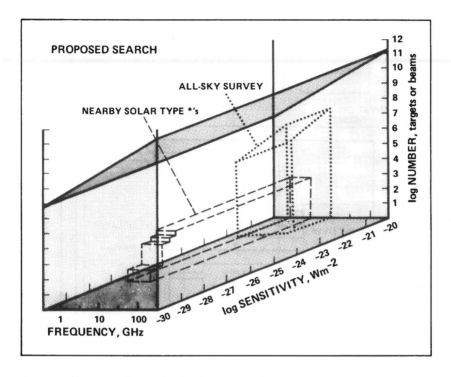

Fig. 7.7. The Cosmic Haystack, showing the search space to be covered by the NASA/ JPL Sky Survey and the NASA Ames Targeted Search. The Targeted Search was designed to have greater sensitivity, while the Sky Survey would observe in more directions and over a broader frequency range. Both were terminated in 1993, but the Targeted Search continued as Project Phoenix with private funding.

survey. The targeted and sky survey strategies were in many ways complementary; only the observations would demonstrate which assumptions were best and which technique was most effective in terms of a successful detection.

The heart of the system and the key to its success was the digital spectrum analyzer. Radio astronomy had never before attempted to use multichannel spectrometers at the scale of the millions of channels needed for the SETI search. The key to the new spectrometer was the advance of digital technology. Beginning with work on a 74,000-channel prototype MCSA in 1977, it took 15 years to arrive at the customized very large integrated circuit (VLSI) chip that became operational at Arecibo in 1992. Another crucial component of the SETI system was the method for extracting an extraterrestrial signal coming through the spectrum analyzer. While detection of signals from noisy data was a standard problem in communications, SETI was a particular challenge because nothing was known with certainty about the nature of an artificial extraterrestrial signal. The signal detection team at Ames therefore had to make a variety of assumptions in designing their signal detection algorithms. In addition, because the system had to to reject any terrestrial radio frequency interference (RFI), the RFI problem was studied extensively and the result implemented in the software algorithms. Finally, because millions of channels were to be analyzed in real time, great demands were placed on the data acquisition system, which had to be specially designed for the project. As these events unfolded at Ames, parallel events took place at JPL.

Administratively, SETI had gone from a few people within a division at Ames in 1976 to two project offices in two centers with a combined staff and subcontractors of about 65 in 1992. Fiscally, its annual budget had risen from a few hundred thousand dollars in the early 1970s to over $10 million in the 1990s. Conceptually, its strategy had been honed and reduced to politically realistic proportions since the visionary Cyclops days. While at NASA headquarters the SETI program had spent most of its lifetime in the Life Sciences Division, in 1992 it became the first element in the Solar System Exploration Division's TOPS (Toward Other Planetary Systems) program designed to detect other planetary systems.

On October 12, 1992, symbolically the quincentennial of Columbus's landfall in the New World, the NASA HRMS was inaugurated amid considerable fanfare. On that date the 305-meter radio telescope at Arecibo began the Ames Targeted Search, while the 34-meter antenna at the Venus station of the Deep Space Communications Complex at Goldstone in the Mohave Desert began the JPL All-Sky Survey. Although these observations were intended to mark only the beginning of an extended enterprise, hopes were soon dashed by Congress. After more than 15 years of planning, $60 million in R&D, and less than a year of observing, the program was terminated as part of congressional budget cuts. SETI, a relatively small program by NASA standards, had often been singled out for scrutiny in the past by Congress, and this time Senator Richard Bryan of Nevada successfully led the opposition to what he called "the Great Martian Chase." "The Great Martian Chase may finally come to an end," Bryan said. "As of today, millions have been spent and we have yet to bag a single little green fellow. Not a single martian has said take me to your leader, and not a single flying saucer has applied for FAA approval," leaving the discerning reader to figure out that SETI had nothing to do with either Martians or flying saucers.

Almost exactly a quarter-century after Billingham had initiated NASA interest in SETI, it was unceremoniously excised from the U.S. government, not only cutting off the federal government's financial and intellectual sponsorship of the country's flagship SETI effort, but also denying the support NASA had given to other projects. A significant number of the project's personnel, however, joined the nonprofit SETI Institute, which had considerable success in raising private funding to carry on at least a scaled-back version of the original Targeted Search. Known as Project Phoenix, it was headed by the NASA SETI veteran Jill Tarter, whose team continued observations at Green Bank and Arecibo Observatory, and at Parkes Observatory in the Southern Hemisphere. And the long-running programs sponsored by the Planetary Society, Ohio State University, and the University of California at Berkeley not only continued with even greater importance, but were upgraded and sporadically joined by new observations.

Thus, although the demise of the NASA program was a severe blow to SETI advocates, and though the political hazards of government funding for SETI had been amply demonstrated, in the last decade of the century SETI remained well entrenched both as a topic for interdisciplinary scientific discussion and as an observational activity. Meetings on extraterrestrial life, which had been gaining momentum for two decades (Table 7.4), not only continued, but were given new life by the discoveries of extrasolar planetary systems and possible fossilized life on Mars, both of which were seen as increasing the chances for intelligent life in the universe. And meetings devoted to SETI (Fig. 7.8) were only a subset of the much larger meetings held to encompass all of the subjects related to the extraterrestrial life debate. Once again, SETI proponents and their exobiological colleagues demonstrated, their subject was too important to be derailed by mere politics.

Thus, modern SETI research continued, and even thrived, despite funding problems and a serious crisis in rationale that undermined the very basis of its observing programs. The challenge to SETI optimists around 1975 marked an important cautionary turning point in its fortunes, which for a time seemed boundless. In the end, however, the Fermi Paradox and all the talk about alternative approaches had little effect on the strategy adopted. NASA and other ongoing programs would not search for Type II or Type III Kardashev or Dyson civilizations, or search for laser communication with civilizations more similar to ours, or (although more directly related to NASA's mission) institute any plans for interstellar travel. In a neat closure to the events of Cocconi, Morrison, and Drake three decades before, NASA and most other searches would undertake a search for Type I civilizations in an expanded region of the electromagnetic spectrum that included the 21-cm line, similar in principle to the earliest thoughts about the opening of the spectrum for SETI, even if now much more powerful. Although there was renewed debate about the merits of optical SETI toward the end of the century, the 21-cm choice "still appeals to most of the radio astronomers who have attended to the problem," Morrison wrote in 1985. "Plainly this is no objective matter, one can only argue plausibility, and then depend upon hope."

Table 7.4. *Selected conferences on extraterrestrial life*

Date	Sponsor	Location	Proceedings
November 1961	U.S. National Academy of Sciences	NRAO Green Bank	Cameron, *Interstellar Communication* (1963)
Feb. 26–28, 1963	Jet Propulsion Laboratory	Pasadena, Calif.	Mamikunian and Briggs, *Current Aspects of Exobiology* (1965)
May 1964	Armenian Academy of Sciences	Byurakan Astrophysical Observatory	Tovmasyan, *Extraterrestrial Civilizations* (1965; English trans., 1967)
1964–1965	U.S. National Academy of Sciences (Space Science Board)	Stanford, Rockefeller Institute	Pittendrigh, Vishniac, and Pearman, *Biology and the Exploration of Mars* (1966)
1970	NASA	Ames Research Center	Ponnamperuma and Cameron, *Interstellar Communication* (1974)
1971	Stanford/NASA	Stanford	*Project Cyclops* (1971)
1971	U.S. and USSR[a] Academy of Sciences	Byurakan	Sagan, *Communication with Extraterrestrial Intelligence* (1973)
1974	American Anthropological Association		Maruyama, *Cultures Beyond the Earth* (1975)
1975–1976	NASA/Ames	Ames, Caltech, Arecibo, Goddard	Morrison, Billingham, and Wolfe, *The Search for Extraterrestrial Intelligence* (1977)
1976	NASA/ASEE	Stanford	Black, *Project Orion* (1980)
1975–1977	International Academy of Astronautics	Lisbon, California	Billingham and Pesek, *Communication with Extraterrestrial Intelligence* (1979)
June 1979	NASA/Ames	Ames Research Center	Billingham, *Life in the Universe* (1980)
August 1979	International Astronomical Union[b]	Montreal	Papagiannis, *Strategies in the Search for Life in the Universe* (1980)
November 1980		University of Maryland	Zuckerman, *Extraterrestrials: Where Are They?* (1981)

Table 7.4. *(cont.)*

Date	Sponsor	Location	Proceedings
December 1981	Soviet Academy[a]	Tallin, Estonia	None published
June 18–21, 1984	International Astronomical Union[b]	Boston, Mass.	Papagiannis, *The Search for Extraterrestrial Life: Recent Developments* (1985)
October 1984	International Academy of Astronautics	Lausanne, Switzerland	
June 22–27, 1987	International Astronomical Union[b]	Balaton, Hungary	Marx, *Bioastronomy: The Next Steps* (1988)
June 18–23, 1990	Third International Bioastronomy Symposium[b]	Val Cenis, France	Heidmann and Klein, *Bioastronomy: The Exploration Broadens* (1991)
August 5–9, 1991	University of California[a] SETI Institute	University of California, Santa Cruz	Shostak, *Third Decennial U.S.–USSR Conference on SETI* (1993)
1991–1992	NASA/ SETI Institute	Santa Cruz	Billingham et al., *Social Implications of Detecting an Extraterrestrial Civilization* (1994)
August 1993	University of California[b] SETI Insitute ISSOL	University of California, Santa Cruz	Shostak, *Progress in the Search for Extraterrestrial Life* (1995)
July 1–5, 1996	Fifth International Bioastronomy Conference[b]	Capri, Italy	Cosmovici et al., *Astronomical and Biochemical Origins* (1997)

[a] U.S.–USSR Decennial SETI Conference series.
[b] Triennial Bioastronomy Conference series, organized by the International Astronomical Union's Commission 51 (Bioastronomy), and others.

It is always difficult to assess the importance of any endeavor from close range, let alone one as far-reaching as SETI, whose importance will depend to a large extent on its success. Clearly, the verdict on modern SETI remained uncertain at century's end. Whether lack of detections, or lack of political will, or an increasingly narrow focus on practical projects with immediate

Fig. 7.8. Participants in the third Decennial U.S.–USSR meeting on SETI at Santa Cruz, California, in 1991. SETI pioneers include (front row, standing) Michael Klein (JPL), Frank Drake (holding N EQLS L, a reference to the Drake Equation), to his right Barney Oliver of Project Cyclops fame, and to his right Jill Tarter, John Billingham, and Carl Sagan. Kneeling below Billingham is the Soviet SETI pioneer Nikolai Kardashev, and Amahl Drake to the left of Kardashev.

benefits, or flagging human curiosity would result in a limited lifespan for modern SETI, and whether a century from now it would be seen as only a curious episode in the history of science like interplanetary communication, only the future will tell. Either way, SETI will surely be seen as the 20th century's effort in the venerable tradition of seeking humanity's place in the universe.

On one thing everyone agreed: if by judicious rationale and strategy or by good fortune, SETI programs defied all the odds and resulted in the discovery of extraterrestrial intelligence, the effect on humanity would be profound. The human implications of that discovery had been a favorite subject of science fiction for decades, and the contemplation of the problem quickly led to philosophy, religion, and questions about purpose in the universe. It is those issues that we shall address in our next chapter.

8

THE MEANING OF LIFE
Implications of Extraterrestrial Intelligence

And now at last the highest truth on this subject remains unsaid; probably cannot be said. . . .

Emerson (1841)

We are in deep waters here, in a sea of great mysteries.

E. A. Milne (1952)

O, be prepared, my soul!
To read the inconceivable, to scan
The million forms of God those stars unroll
When, in our turn, we show to them a Man.

Alice Meynell (1923)

Never before have so many been so enthusiastic about being so trivial.

Rabbi Norman Lamm (1978)

At the end of our history, we return to the question with which we began: the meaning of extraterrestrial life for humanity. We have argued in a general sense that belief in extraterrestrial life is a kind of worldview in itself, a "biophysical cosmology" that – once accepted – forever changes the perception of our place in the universe. We have also seen (Chapter 1) that the discussion of the mere *possibility* of extraterrestrial life has in prior centuries given rise to a great diversity of opinion on the possible meanings for humanity, particularly in the religious context.

In this chapter we examine how 20th-century thinkers from diverse backgrounds have viewed the implications of life beyond the Earth for human culture, theology, and philosophy. In exploring humanity's relation to other intelligent species, these thinkers have, however tentatively, extended the bounds of philosophy and theology beyond their usual anthropocentric perspectives, themselves becoming pioneers whose ideas remain for the most part undeveloped. Unlike the scientific issues in the debate, the question of implications has received very little systematic study for the simple reason that the humanities and social sciences have not yet taken to heart the potential implications of the biological universe, having had quite enough to contend with in studying the sometimes inexplicable activities of the local intelligent species. The question remains a subject for the future, one that may draw on the sporadic and widely divergent results that the history of life on Earth has offered thus far.

8.1 PERCEPTIONS OF CULTURAL IMPACT

Though the subject had been occasionally addressed earlier, the Space Age gave immediacy to the problem of the impact of a discovery of extraterrestrial life. Already in 1961, a NASA-sponsored study undertaken by the Brookings Institution, in compliance with the National Aeronautics and Space Act to identify long-range goals of the U.S. space program and their effect on American society, offered a discussion of the implications of a discovery of extraterrestrial life. Its broad-based report touched only briefly on life in space, but in doing so raised a variety of important questions. The social science authors viewed the recently completed Project Ozma as having popularized and legitimized speculations about the impact of such a discovery on human values; it also led them to agree with their astronomical colleagues that radio signals seemed the most likely mode of contact over the next 20 years.

The NASA/Brookings report emphasized that both individual and government reactions to radio contact with intelligence would likely depend on religious, cultural, and social backgrounds, as well as on any information communicated. The mere knowledge of life in the universe, they speculated, could lead to greater unity on Earth based on the uniqueness of humanity or its unified reaction to something alien. Because of the difficulties in communication, a discovery could become "one of the facts of life" not requiring any action. On the other hand, in a statement often cited since, the authors also warned that substantial contact could trigger a more foreboding effect: "Anthropological files contain many examples of societies, sure of their place in the universe, which have disintegrated when they had to associate with previously unfamiliar societies espousing different ideas and different life ways; others that survived such an experience usually did so by paying the price of changes in values and attitudes and behavior." Scientists, they suggested, might be the most devastated of all by the discovery of superior intelligence, since "an advanced understanding of nature might vitiate all our theories at the very least, if not also require a culture and perhaps a brain inaccessible to earth scientists." Finally, the authors pointed out that philosophical problems could be raised, particularly in deciding whether aliens were to be treated morally and ethically as human beings.

The NASA/Brookings study set forth in broad outline the alternative impacts of extraterrestrial intelligence: good or bad and possibly worst of all for scientists. As for the discovery of plant life or "subhuman intelligence" on Mars or Venus – the planets soon to be explored by the space program – they proposed that after the initial novelty had worn off, large parts of the American public might not be affected any more than they were by the discovery of the coelacanth or the panda.

Members of the Space Science Board of the National Academy of Sciences, however, in reviewing the scope of space biology while contemplating the first Mars missions in 1962, were not so sanguine about the philosophical import of exobiology, even when dealing with low forms of life: "It is not since Darwin," they wrote, "and before him Copernicus – that science has had the opportunity for so great an impact on man's understanding of man. The scientific question at stake in exobiology is, in the opinion of many, the most exciting, challenging, and profound issue, not only of this century but of the whole naturalistic movement that has characterized the history of Western thought for three hundred years. What is at stake is the chance to gain a new perspective on man's place in nature, a new level of discussion on the meaning and nature of life." In making this statement, the Space Science Board was not merely expressing an academic opinion: its members not only foresaw significant implications of a discovery of extraterrestrial life, they also used their positive conception of these implications to argue for an exobiology program of considerable expense to the public.

The ability to understand a more generalized biology, however, was quite different from a concern about the implications of the discovery of intelligence. It was the latter that continued to intrigue most academics drawn to the subject and that was the focus of a 1-day symposium on "Life Beyond Earth and the Mind of Man" sponsored jointly by NASA and Boston University in 1972 (Fig. 8.1). Billed by its moderator, the Boston University astronomy professor Richard Berenzden, as the first meeting in which a distinguished panel from diverse fields discussed in an open forum the ramifications of a discovery of extraterrestrial intelligence, the symposium produced some provocative (but hardly systematic) ideas and is perhaps most notable for its diversity of opinion rather than for any consensus. The Nobel Prize-winning Harvard biologist George Wald, allowing that the universe was probably full of life and that long-distance communication was more likely than physical contact, declared that he could "conceive of no nightmare as terrifying as establishing such communication with a so-called superior (or if you wish, advanced) technology in outer space." Though some gleefully prophesy great benefits from such communication, Wald continued – conjuring the scenario already broached by the Brookings report – "the thought that we might attach, as by an umbilical cord, to some more advanced civilization, with its more advanced science and technology, in outer space does not thrill me, but just the opposite." For humanity itself to discover a cure for cancer, or to control nuclear fusion, was one thing, he said. "But just to get such information passively from outer space through that transmission is altogether different. One could fold the whole human enterprise – the arts, literature, science, the dignity, the worth, the meaning of man – and we would just be attached as by an umbilical cord to that 'thing out there'."

Fig. 8.1. Boston University symposium on Life Beyond Earth and the Mind of Man, 1972. From left to right: George Wald, Philip Morrison, Richard Berenzden, Carl Sagan, Krister Stendahl, Ashley Montagu. Boston University Photo Services.

Wald's fears found little sympathy from other participants. The theologian Krister Stendahl, dean of the Harvard School of Divinity, viewed the growing awareness of extraterrestrial intelligence as not at all threatening. Rather, he believed such a discovery would raise cosmic consciousness, make God's universe even better, and leave humanity "relatively unique" rather than "absolutely unique" in the universe. Notably, however, Stendahl raised no discussion of dogma. The three astronomers present – Morrison, Sagan, and Berenzden – all found Wald's fears groundless, as did the anthropologist Ashley Montagu.

In a more general sense, Morrison argued that the discovery of intelligence and its subsequent impact would not be a quick event, but would resemble more the development of agriculture than the discovery of America. Reaction to even the latter discovery, he pointed out, was slow. He believed the message would not be simple; the signal would be immediately recognizable, but the message content would be only slowly deciphered. "It will be more of a subtle, long lasting, complex, debatable effect than a sudden revelation of truth, like letters written in fire in the sky," he said. Morrison was "neither fearful nor terribly expectant. I am anxious for that first acquisition, to make sure that we are not alone. But once that is gained – it might be gained in my lifetime – then I think we can rest with some patience to see what complexities have turned up on other planets."

Also in a positive vein, Sagan argued that discovering extraterrestrial intelligence would reestablish a context for humanity that had been long lost. "The old secure sense of where we are in the universe has eroded. . . . The kind of exploratory ventures we are talking about seem to me to be precisely the kind that are needed to reestablish a cosmic context for mankind. By finding out what the other planets are like – by finding out whether there are civilizations on planets of other stars – we reestablish a context for ourselves."

Rejecting the likelihood of invasion, exploitation, or subversion of the human race by extraterrestrials, Berenzden agreed that the discovery of intelligence beyond the Earth would give us access to our "galactic heritage," not merely in terms of technology and science, but also in the arts, literature, and humanities. The culture shock from radio contact, Berenzden believed, was problematic. The first contact would likely be only a confirmation of existence, followed "many hundreds of years later" by information trickling in, and then by tens or hundreds of years of decipherment.

Neither Berenzden nor the other participants could begin to predict the impact of extraterrestrial contact on humanity. But NASA Administrator James Fletcher, noting that technically advanced civilizations in other planetary systems were not only possible but likely, urged that the provocative ideas of the six panelists were "worthy of consideration by thoughtful persons everywhere." In calling for more study, Fletcher echoed the Brookings report, which also recommended continued studies of the attitudes toward the possibility of extraterrestrial intelligence and its discovery, as well as historical and empirical studies of human behavior when confronted with dramatic and unfamiliar events or social pressures. The Brookings authors believed such studies would illuminate how the news of a discovery should be presented to or withheld from the public, as well as the role of scientists and government in releasing this information.

That NASA had a hand in all three of these studies – the Brookings report, the Space Science Board meeting, and the Boston discussion panel – is significant; the broader intellectual community was still reticent to raise the subject without some prodding. One exception was the philosopher Lewis W. Beck, who echoed the call for further study and offered two conjectures of his own after reviewing the history of the debate. First, he argued, popular science and science fiction have so prepared the public for the idea of signals from space that such a discovery will be forgotten after a few weeks, just as the details of the Moon landing were forgotten by the majority of people, who had more immediate concerns in their daily lives. But he also conjectured that, in a larger sense, the discovery of intelligence beyond the Earth would never be forgotten: "For what is important is not a single discovery, but the beginning of an endless series of discoveries which will change everything in unforeseeable ways." If further discoveries are made, "there is no limit to what in coming centuries we might learn about other creatures and, more portentously, about ourselves. Compared to such advances in knowledge, the Copernican and Darwinian Revolutions and the discovery of the New World would have been but minor preludes."

By the early 1970s, then, there were many calls for further study but no consensus among the few groups that had discussed, all under NASA sponsorship and on a very small scale, the potential impact of contact. Nor, Beck aside, was there any inclination among social scientists or philosophers to address the problem of extraterrestrial implications independently and no answer to the call for more study during the 1970s.

On the other hand, both inclination and near consensus did exist among one group – the SETI participants themselves. These began with neutral declarations of significance: Cocconi and Morrison, in their pioneering article in 1959, wrote that "Few will deny the profound importance, practical and philosophical, which the detection of interstellar communications would have," while in the year of Project Ozma, Drake closed one of his articles on the subject with the words "the scientific and philosophical implications of such a discovery will be extremely great." But these neutral statements quickly took on an air of advocacy. The Introduction to the earliest anthology of papers on interstellar communication – many of them based on the 1961 Green Bank meeting – commented that in admitting that millions of advanced societies existed in our Galaxy, we were completing the Copernican revolution. Moreover, "If we now take the next step and communicate with some of these societies, then we can expect to obtain an enormous enrichment of all phases of our sciences and arts. Perhaps we shall also receive valuable lessons in the techniques of stable world government," *Interstellar Communication* editor A. G. W. Cameron commented at the height of the Cold War.

This brief comment reflected the attitude of most other SETI researchers, an attitude that by the time of the Boston symposium had become widespread. The Project Cyclops report, whose influence in this area, as in its technical treatment of SETI, cannot be underestimated, was particularly important in this regard. In this report Bernard Oliver had spoken of "our galactic heritage" and "the salvation of the human race" in the context of extraterrestrials while downplaying possible hazards such as invasion, exploitation, subversion, and culture shock. Galactic civilizations are likely to have first formed 4 or 5 billion years ago, Oliver pointed out, and thus interstellar communication may have been ongoing for a long time. The "cosmic archaeological record of our Galaxy," Oliver argued, might lead to our self-preservation, to new branches of science, to the end of the cultural isolation of the human race, and to a re-orientation of our philosophy. While not totally lacking risk, Oliver argued, the culture shock exhibited repeatedly by culture contact on Earth was unlikely to be repeated via radio contact. Rather, the "long delays and initially slow information rate" would allow us to adapt to the new situation.

Philip Morrison espoused a similar view almost simultaneously at the joint U.S.–USSR Byurakan meeting in 1971. The complex signal arriving at our radio telescope, he argued, "is the object of intense socially required study for a long period of time. I regard it as much more like the enterprise of history of science than like the enterprise of reading an ordinary message. . . . The data rate will for a long time exceed our ability to interpret it." This scenario of gradual assimilation, he asserted, directly illuminates the nature of the impact on humanity: "The recognition of the signal is the great event, but the interpretation of the signal will be a social task comparable to that of a very large discipline, or branch of learning. In that light, I think that a sober study will show that a message channel cannot open us to the sort of impact which we have often seen in history once contact is opened between two societies at very different levels of advance. . . . We could imagine the signal to have great impact – but slowly and soberly mediated."

Frank Drake was more effusive about immediate impact. Contact, especially with the "immortals," could give us a shortcut to wisdom, he believed. Not only would we gain scientific and technical information, we might also learn about "ultimate social systems," art forms, and other aspects of life as yet unimagined. If this is so, he concluded, "let us know it now. We need not be afraid of interstellar contact, for unlike the primitive civilizations on earth which came in contact with more advanced technological societies, we would not be forced to obey – we would only receive information." Both Drake and Sagan believed such a discovery "would inevitably enrich mankind beyond measure."

These attitudes were in the background of the Morrison workshops of 1975–1976, which, while most notable as a technical landmark in converting SETI from discussions into a realistic program, also offered some thoughts on the impact of SETI. A single extraterrestrial signal, the participants noted in *The Search for Extraterrestrial Intelligence,* would lead immediately to one great truth: "that it is possible for a civilization to maintain an advanced technological state and *not* destroy itself." Pointing out that humanity was under no obligation to reply to a radio message, and could even ignore one that seemed offensive, the Morrison report rejected negative impacts and argued that a signal would "pose few dangers to mankind; instead it holds promise of philosophical and perhaps practical benefits for all of humanity."

In contemplating an actual SETI search, the report also broached another issue that had heretofore received little attention: that early claims of signal detection may turn out to be mistaken; that the press and the public must use caution to avoid raising hopes and fears, and thus "the importance of a skeptical stance and the need for verification." Beyond the acquisition of the first genuine signal, the report echoed Morrison's previously stated attitude that study of any information content would occupy generations. The commonality of radiophysics and mathematics, it stressed, might allow a signal to be decoded, but perhaps only "in a slow and halting manner." Still, the report urged that the question of impact "deserves rather the serious and prolonged attention of many professionals from a wide range of disciplines – anthropologists, artists, lawyers, politicians, philosophers, theologians – even more than that, the concern of all thoughtful persons, whether specialists or not." One of those professionals, Theodore Hesburgh, the president of Notre Dame University, declared in the Foreword to the report that SETI and theology were not incompatible. To the contrary, Hesburgh held, echoing 18th-century natural theology arguments: "As a theologian, I would say that this proposed search for extraterrestrial intelligence (SETI) is also a search of knowing and understanding God through His works – especially those works that most reflect Him. Finding others than ourselves would mean knowing Him better."

Not everyone agreed with the rosy view of impact offered by the Morrison report. In particular, British Astronomer Royal and Nobelist Sir Martin Ryle fell in to the Wald camp. After reading of Drake's attempt to signal (as opposed to listen to) extraterrestrial intelligence as part of the dedication of the newly resurfaced Arecibo dish, Ryle addressed an appeal to the International Astronomical Union urging that no attempts be made to signal other civilizations. *The New York Times,* in a 1976 editorial entitled "Should Mankind Hide?", opted for the optimistic side, concluding that "there is no reason to assume that alien intelligence among the stars must be hostile or predatory." Neither, of course, was there any reason to assume that it would be benevolent, nor did they seem worried by Wald's thesis of umbilical subservience.

During the 1980s no further consensus was reached on the implications of contact with extraterrestrials, nor perhaps could this have been expected. SETI enthusiasts such as Sagan took their message of hope to an ever-larger population. Sagan's *Cosmos* (1980) spread the message that discovering another civilization would be "a profoundly hopeful sign. . . . It means that someone has learned to live with high technology; that it is possible to survive technological adolescence." The evolutionary paleontologist Stephen Jay Gould argued, without committing to beneficial or catastrophic implications, that "A positive result [of SETI] would be the most cataclysmic event in our entire intellectual history." But others, notably the philosopher Edward Regis, Jr., argued against profound consequences. If a signal was intercepted, he claimed, "it is probably as likely to be wholly unintelligible as it is that it will bring us the answer to life, the universe, and everything." The resistance to decryption of even some earthly texts, in his view, did not inspire confidence that a truly alien message would be decipherable.

The 1980s and 1990s did see the beginning of systematic discussion, if not resolution, of some of the problems raised in the previous two decades. More tractable than the long-term impact, however, and increasingly important as search programs such as NASA's neared reality, was the problem of the receipt, verification, and announcement of a signal, which came to be subsumed under the subject of cultural impact, since the manner in which the announcement was made could well determine the impact. Considerable energy was spent on what came to be known as the "postdetection protocols," a discussion in which NASA and the International Academy of Astronautics again took the lead. Building on work begun a few years earlier, in 1986 and 1987 the International Academy of Astronautics and the International Institute of Space Law cosponsored a series of meetings during the annual congress of the International Astronautical Federation, where SETI had been a topic of more technical discussions for two decades. The hope of these meetings was "to uncover particular principles or guidelines that should be applied to the conduct of SETI programs and the communication of their extraordinary important results." The discussions also went beyond that goal to discuss diplomatic and legal issues relating to contact and communication. In the end, these efforts succeeded in drafting postdetection protocols in the aftermath of a message, but they did not address the more difficult question of response. And only 2 of the 16 papers at the International Astronautical Federation meetings dealt with the thornier problem of impact after a message was received; the SETI Committee of the International Academy of Astronautics, however, continued to wrestle with these problems.

Whereas the International Astronautical Federation meetings were inspired by a desire to discuss the issues of near-term receipt, verification, and announcement of a signal, in 1990 more systematic studies began to examine the entire scope of short- and long-term implications. In particular, under the direction of John Billingham and on the eve of the launching of its own observations, in 1990–1991 NASA sponsored a series of three workshops on Cultural Aspects of SETI (CASETI) to discuss the cultural, social, and political consequences of a successful detection of extraterrestrial intelligence. Toward this end, a variety of about 25 social and physical science specialists addressed the relevance of four broad areas to such a detection: history, behavioral science, national and international policy, and education. Dividing the historical viewpoints into "millenarian" and "catastrophist" categories, the CASETI historians found that history could offer useful, though not definitive, analogues of the discovery of intelligence. In this respect they recommended that further study be given not to physical culture contacts on Earth, but to intellectual contacts between cultures, as for example when Greek science and learning was passed via the Arab culture to the Latin West in the 13th century. Offering their own assessment of impact, the historians wrote that "it takes a very dour catastrophist to believe that mere knowledge of the existence of ETI could extinguish the manifold creative energies that have enabled mankind to overcome many sorts of menaces over the centuries." The behavioral scientists, however, emphasized the extremely diverse reactions likely to take place among many social and religious groups, and cautioned that predicting any reaction was an extremely complex endeavor, even if mass education prior to detection attempted to precondition responses. This did not preclude the recommendations of the education specialists, who urged that it be carried out in the broadest possible manner. And the policy specialists, looking beyond the work of the International Academy of Astronautics, encouraged broader institutional policies beyond the postdetection protocols already envisaged.

It seemed likely that by the end of the century the discussions of impact, so tentatively begun with the dawn of the Space Age, would rise to a new level of sophistication. Indeed, this is just what happened with the publication in 1997 of Albert Harrison's *After Contact: The Human Response to Extraterrestrial Life,* which used "living systems theory" to discuss human reactions based on the fields of psychology, sociology, anthropology, and political science. The concrete achievement of the postdetection protocols also seemed likely to be expanded and the ideas of longer-term impact debated. While consensus on the latter seemed unlikely, many participants undoubtedly shared the opinion of the anthropologist Ashley Montagu, who declared the importance of preparation for extraterrestrial contact, no matter how unlikely the prospect of detection or communication: "I do not think we should wait until the encounter occurs," he urged. "We should do all in our power to prepare ourselves for it. The manner in which we first meet may determine the character of all our subsequent relations."

8.2 ASTROTHEOLOGY

The question of the impact of extraterrestrial life on religion and theology has very deep roots, at least in the Western tradition. The problem was perceived in the 15th century in relation to the reconciliation of Christianity with the Aristotelian doctrine opposing a plurality of worlds. Most theologians by that time agreed that God could create other worlds. But if so, they wondered whether Christ's death on this earth could redeem the inhabitants of another world. The standard answer was that he could because Christ could not die again in another world. Very early in the Protestant tradition, Martin Luther's supporter, Philip Melanchthon, not only objected to such a speculative idea but also used it as an argument against the Copernican theory. "It must not be imagined that there are many worlds, because it must not be imagined that Christ died and was resurrected more often, nor must it be thought that in any other world without the knowledge of the Son of God, that men would be restored to eternal life." For Copernicans of any religious persuasion, the problem was a thorny one that extended beyond specific religious doctrine. Kepler stated the conundrum in the early 17th century in more general terms that might equally apply to other religions of the world: "If there are globes in the heavens similar to our earth, do we vie with them over who occupies a better portion of the universe? For if their globes are nobler, we are not the noblest of rational creatures. Then how can all things be for man's sake. How can we be the masters of God's handiwork?"

These provocative Keplerian questions were still alive at the end of the 19th century, when H. G. Wells quoted them as the prelude to his novel *War of the Worlds*. By that time, as we have seen in Chapter 1, Christianity had explored these implications quite substantially. Despite Scriptural objections raised during the 17th century, by the early 18th century the Anglican priest and Royal Society Fellow William Derham reflected accepted theological opinion when he incorporated extraterrestrial life into natural theology; it is in the sense of inhabited worlds reflecting the magnificence of God's universe that Derham wrote his book *Astro-Theology*. The matter did not rest there, however. Thomas Paine bluntly stated in *Age of Reason* (1793) that extraterrestrials and Christianity did not mix, and that "he who thinks that he believes in both has thought but little of either." In a history that would repay study by those interested in theological implications of an actual discovery of extraterrestrial intelligence, during the 19th century some writers rejected Christianity, others rejected plurality of worlds, and still others found ways to reconcile the two.

The 20th century thus inherited a considerable discussion of the theological implications of extraterrestrial life, mostly within the Christian tradition, inspired by the mere *possibility* of intelligence beyond the Earth. Although the relation between theology and plurality of worlds occasionally reached the level of sustained debate in the 18th and 19th centuries, by the mid-20th century this controversy echoed only faintly in the background as scientists began to contemplate the possibility of a search for extraterrestrial intelligence. In the 20th century Derham's "astrotheology" assumed new meaning in light of efforts to detect signals from extraterrestrial intelligence, efforts that, if successful, would surely affect traditional theology, with its emphasis on the relation between God and humanity. Rather than focusing on *confirming* evidence of the glory of God in the best tradition of natural theology, astrotheology in the 20th century came to describe the considerable *modifications* of theology and religion that might develop in the wake of the discovery of intelligence in the heavens. It is this latter tradition that we trace in this section.

During the 20th century, theological explorations of contact with extraterrestrial intelligence have been sporadic, peripheral, and never raised to the same level of substantial debate achieved in the previous two centuries. Indeed, lacking a more explicit discussion of the problem by practitioners, the attitudes of many religions have had to be deduced by others from the general doctrines of those religions. The major exception has been Christianity, where the doctrine of Incarnation has been a central focus of discussion and where the consensus has been that a discovery of intelligence beyond the Earth would not prove fatal to the religion or its theology. In general, for Christians as well as for other religions, indigenous theologians see little problem, while those external to religion proclaim the fatal impact of extraterrestrials on Earth-bound theologies.

While most religions would undoubtedly have preferred to remain silent on the subject, the issue was pushed into the public and theological consciousness by the approach of the Space Age. As Arthur C. Clarke, one of the prophets of the new era, remarked in his popular book *The Exploration of Space* (1951), some people "are afraid that the crossing of space, and above all contact with intelligent but nonhuman races, may destroy the foundations of their religious faith. They may be right, but in any event their attitude is one which does not bear logical examination – for a faith which cannot survive collision with the truth is not worth many regrets." Religion could not for long avoid such a commonsense challenge, whose force could only increase as rocketry neared reality.

Christian flexibility is evident in the earliest 20th-century discussions of the subject, centered at Oxford University, where (not coincidentally) C. S. Lewis had written his novels of Christian apologetics in a science fiction setting. The first Oxford contribution came from a cosmologist's attempt to reconcile the modern view of the universe with Christianity. In one of a series of essays written just days before his sudden death in 1950, Oxford Professor of Mathematics Edward A. Milne took up the problem in connection with the second law of thermodynamics. Milne concluded that the universe would not end in a "heat death" as entropy reached a maximum; rather, it would continue to exist with an infinite variety of forms. What, then, was the fate of this infinite variety? Recalling Eddington's statement that nature might require millions of acorns to grow a single oak, and his implication that an infinity of galaxies might be needed to produce a single planetary system and an infinity of planetary systems to produce one with life, Milne found another possibility more attractive. "The infinity of galaxies," he wrote, could be considered "as an infinite number of scenes of experiment in biological evolution," a view Milne found more appropriate to an omnipotent God. Not only did Milne's God endow the universe "with the only law of inorganic nature consistent with its content," he also "tended his creation in guiding its subsequent organic evolution on an infinite number of occasions in an infinite number of spatial regions. That is of the essence of Christianity, that God actually intervenes in History."

But, Milne noted, this raised a difficulty "which many Christians have felt" with regard to the most notable case of God's intervention in history – the Incarnation: "Was this a unique event, or has it been re-enacted on each of a countless number of planets? The Christian would recoil in horror from such a conclusion. We cannot imagine the Son of God suffering vicariously on each of a myriad of planets. The Christian would avoid this conclusion by the definite supposition that our planet is in fact unique." Milne himself was not satisfied with this supposition, and seeking a way to reconcile Incarnation and life in the universe, he found an answer in interstellar communication, one of the earliest references to the subject. Pointing to the new science of radio astronomy and the reception of radio signals from sources in the Milky Way announced two decades earlier, Milne suggested, "It is not outside the bounds of possibility that these are genuine signals from intelligent beings on other 'planets', and that in principle, in the unending future vistas of time, communication may be set up with these distant beings." In that case, Milne continued, even groups of galaxies would "by inter-communication become one system." Therefore "there would be no difficulty in the uniqueness of the historical event of the Incarnation. For knowledge of it would be capable of being transmitted by signals to other planets and the re-enactment of the tragedy of the crucifixion in other planets would be unnecessary." Were Milne alive today, therefore, he would undoubtedly champion SETI as a means of interplanetary salvation.

The second Oxford volley came only 4 years later, when Milne's clerical contemporary at Oxford, E. L. Mascall, agreed that Christianity could be flexible on the issue but disagreed with Milne's solution. As part of a broad-ranging discussion of the relation of science and theology, Mascall, an Anglican priest and lecturer in philosophy of religion, elaborated Milne's discussion as an example of how modern science could stimulate and expand the scope of theological discussion. Mere knowledge of terrestrial Incarnation brought about by interstellar radio communication was, in Mascall's view, not enough for universal salvation as conceived in classical Christian thought. The latter, he believed, required a "hypostatic union" between Redeemer and redeemed, which Mascall viewed as possible on other planets, no less than on Earth: "There are no conclusive *theological* reasons for rejecting the notion that, if there are, in some other part or parts of the universe than our own, rational corporeal beings who have sinned and are in need of redemption, for those beings and for their salvation the Son of God has united (or one day will unite) to his divine Person their nature, as he has united to it ours. . . ." Mascall went on to more theological technical points – of interest, one would imagine, only to theologians – and probably an indication of the kind of discussion that would follow if SETI were successful. Though he disagreed with Milne on the details, the important point was theological flexibility, and in concluding, Mascall emphasized "how wide is the liberty that Christian orthodoxy leaves to intellectual speculation."

The Oxford excursions into speculative theology, brief though they were, would not be entirely forgotten in subsequent decades. Neither, however, were theologians especially anxious to encourage this kind of speculation. In offering his own view that Incarnation "is unique for the special group in which it happens, but it is not unique in the sense that other singular incarnations for other unique worlds are excluded," the Protestant theologian Paul Tillich characterized the question as one that "has been carefully avoided by many traditional theologians, even though it is consciously or unconsciously alive for most contemporary people."

Nevertheless, the dawning of the Space Age pressed reluctant theologians even further. By 1958 C. S. Lewis, now willing to transport his ideas from fiction to nonfiction, downplayed the effect of extraterrestrials on Christianity, holding that such a discovery would do no more to disprove its principles than did the ideas of Copernicus, Darwin, or Freud. The problem of the Incarnation, he believed, "could become formidable," but only if a variety of conditions were met. Not only would intelligence have to be discovered beyond the Earth, it would also have to possess "rational souls," and it would have to have fallen from grace. If all of this were known to be true, then – and only then – do we need worry about the mode of Redemption – whether it could take place on other planets or whether Christ, by dying on Earth, had already saved extraterrestrials. Lewis, professing to be skeptical about the possibility of intelligence on other worlds, concluded that "a Christian is sitting pretty if his faith never encounters more formidable difficulties than these conjectural phantoms."

Other theologians, however, felt the need to tackle the problem more squarely, perhaps in response to popular interest. The Catholic version of Christianity, like the Protestant, was remarkably open-minded on the subject. Father Daniel C. Raible was typical of this open-mindedness when he wrote in the wake of Project Ozma, "Yes, it would be possible for the Second Person of the Blessed Trinity to become a member of more than one human race. There is nothing at all repugnant in the idea of the same Divine Person taking on the nature of many human races. Conceivably, we may learn in heaven that there have been not one incarnation of God's son but many." The Catholic Church also had an eye on history; quoting a cardinal that "one Galileo case is quite enough in the history of the Church," an editorial in one Catholic journal suggested that "today's theologians would welcome the implications that such a discovery might open – a vision of cosmic piety and the Noosphere even beyond that of a Teilhard de Chardin."

249

The most substantial theological discussion of the subject, and the closest the Roman Catholic Church came to an official position, was given by the priest Kenneth Delano in his book *Many Worlds, One God* (1977). Complete with the official "nihil obstat" and "imprimatur" sanctions, the author's position was that any person with a religious faith, including "an adequate idea of the greatness of God's creative ability, of humanity's humble position in the universe, and of the limitless love and care God has for all His intelligent creatures," should not be afraid to examine the implications of intelligence in the universe. Delano characterized the fears of some in the religious community with regard to extraterrestrials as analogous to early Church skepticism that any humans could live in the terrestrial "antipodes" because none of Adam's descendants could have reached the Southern Hemisphere. Reacting to an early 20th-century writer who claimed that "If he [man] is not the greatest, the grandest, the most important of created things, the one to whom all else is made to contribute, then the Bible writers have misrepresented entirely man's relation to God and the universe," Delano pointed out that God was not obliged to reveal extraterrestrials in the Bible when it would have served no moral purpose. The Space Age requires a theology that is neither geocentric nor anthropomorphic, he argued, and it follows that the Earth may not be the only planet that has seen an Incarnation: "Any one or all three Divine Persons of the Holy Trinity may have chosen to become incarnated on one or more of the other inhabited worlds in the universe." This he considered much more likely than a theory of the "cosmic Adam," in which the single redemptive act by Christ on Earth is applicable to the entire universe. On the other hand, humanity's "mission" could be to spread the Gospel among the inhabited planets while refraining from any form of religious imperialism. The Church, while spreading the story of terrestrial redemption, might also encourage fallen races to seek salvation. In Delano's view, Meynell's poetic vision of the universe was edifying

> . . . in the eternities
> Doubtless we shall compare together, hear
> A million alien Gospels; in what guise
> He trod the Pleiades, the Lyre, the Bear.

Although Delano made it clear that Catholic opinion was not unanimous, he certainly reinforced the prevalent idea of flexibility toward a discovery of extraterrestrials in the Church doctrines.

The same flexibility was expressed in a study of religious implications of the problem for Jewish thought, where the primary concern was, of course, not the Incarnation, but the uniqueness of humanity and its relationship to God. Cautioning that extraterrestrial intelligence was far from proven, Rabbi

Norman Lamm nevertheless pointed to precedents in medieval Jewish thought and declared that in the spirit of open-mindedness toward new knowledge, it was prudent to explore "a Jewish exotheology, an authentic Jewish view of God and man in a universe in which man is not the only intelligent resident, and perhaps inferior to many other races." Medieval Jewish philosophy had already rejected the uniqueness of humanity, Lamm pointed out, but nonsingularity of humanity did not mean insignificance. Harvard astronomer Harlow Shapley and others, he argued, were "profoundly mistaken" in assuming that the number of intelligent species had any relation to the significance of humanity, and even more so in holding that a peripheral position in the Galaxy implied metaphysical marginality and irrelevance. That "geography determines metaphysics" he called a "medieval bias" that should have disappeared with the collapse of geocentrism. Judaism, therefore, "could very well accept a scientific finding that man is not the only intelligent and biospiritual resident in God's world" as long as the insignificance of man was not an accompanying conclusion. Humanity could still be considered unique in "spiritual dignity," and the existence of innumerable intelligences does not lessen God's attention to humans. "A God who can exercise providence over one billion earthmen," Lamm concluded, "can do so for ten billion times that number of creatures throughout the universe. He is not troubled, one ought grant, by problems in communications, engineering, or the complexities of cosmic cybernetics. God is infinite, and He has an infinite amount of love and concern to extend to each and every one of his creatures."

Internal to various religions, therefore, was the consensus that terrestrial religions would adjust to extraterrestrials, an opinion echoed in a recent study of religious attitudes. And, as the same study also pointed out, if the "Adamist religions" of Judaism, Christianity, and Islam – those that share a view of the creation of humanity that links it directly to the godhead – can survive extraterrestrials, then non-Adamist religions such as Buddhism, Hinduism, or the Chinese should have no trouble. In fact, there is already some indication that this would be the case.

This flexible attitude of the world's religions toward extraterrestrials, however, was not shared by some external to those religions. In particular, this was the view taken in perhaps the most detailed discussion of the impact of extraterrestrials on theology by a nontheologian, the philosopher Roland Puccetti. Writing in 1968 in the broad philosophical context of the concepts of "persons" and "moral agents in the universe," Puccetti argued in *Persons* that the world's religions would not be able to survive the implications of an inhabited cosmos. He believed monotheistic religions in particular, involving a personal relationship between a Creator and its creatures, would be in trouble. It is not only that "the prospect of extraterrestrial intelligence, concerning which the principal sacred writings of Christianity, Judaism, and

Islam are absolutely silent, generates a profound suspicion that these terrestrial faiths are no more than that." An even graver problem, in Puccetti's
view, was the "particularism" of these religions, especially when it comes to
doctrines like the Incarnation. Referring to the Milne–Mascall positions regarding that doctrine, he found neither satisfying, Milne's because the size of
the universe precluded spreading the news by radio communication to all potential planets, and Mascall's because his proposed "hypostatic union" led to
multiple divine beings in the universe.

For the Oriental faiths, Pucetti believed that such a discovery would be
a far less serious matter from the point of view of doctrine, since "in their
'higher' forms at least they teach salvation through individual enlightenment
and conceive the supreme Reality in strictly impersonal terms." Even they,
however, suffer from a characteristic "that may be said to affect all terrestrial
religions without distinction." This is what Puccetti terms "particularism,"
those traits that make each religion specific not only to the Earth, but even to
particular cultures on the Earth. For example, Confucianism and Taoism are
deeply immersed in Chinese culture, the former bound up with social, political, and familial duties and the latter with Chinese superstitions. Hinduism
has its caste system and regard for sacred cows; Shintoism holds that the first
divine creation was the islands of Japan; Judaism is intimately tied to a particular part of the human race, the "chosen people"; Islam has its Mecca and
its feasts tied to the Moon; and Christianity, of course, has the greatest particularism in its emphasis on Jesus Christ and the Incarnation. Particularism,
Puccetti argued, has two problems: first, because of this trait, no one religion
can claim to be universal; second, if there are multiple independent religions
on each potentially inhabited planet, then there are billions of independent
claims to truth, surely undermining any concept of truth one would wish to
offer. And if one decides to take the easy way out and deny extraterrestrials, Puccetti argued, this makes terrestrial faiths falsifiable, something they
surely wished to avoid as well. From these considerations, Puccetti concluded,
the existence of extraterrestrials causes us to abandon all particularistic religions, and thus all terrestrial religions, since none of them may be considered
universal.

The dichotomy between internal and external attitudes toward the impact
on religion and theology is highlighted by the response to Puccetii. The Notre
Dame University philosopher Ernan McMullin rejected Puccetti's claim that
the Christian doctrine of Incarnation made no sense in an inhabited universe.
His response confirmed the now characteristic attitude of flexibility adopted
by many religions toward extraterrestrials: "There is an odd, ungenerous fun-
damentalism at work here, a refusal to allow for the expansion of concept,
the development of doctrine, that is after all characteristic of both science
and theology." The most important thing about Puccetti's book, McMullin

concluded, was that it drew attention to the need for more study of theology and philosophy in the context of new astronomical discoveries, including the possibility of extraterrestrial intelligence.

In retrospect, the internal religious response to extraterrestrial intelligence is perhaps not surprising, considering that the alternative to adaptation is destruction. While maintaining flexibility, these religions did not underestimate the challenge. As Delano said, "To build up an immunity to the possibility of a shock that could put an end to, or at least cripple, the practice of religion, we should all do well to exercise the virtue of humility in evaluating our standing in God's creation." This sentiment was echoed by McMullin, who noted that "a religion which is unable to find a place for extraterrestrial persons in its view of the relations of God and the universe might find it difficult to command terrestrial assent in days to come." Still, McMullin admitted in 1980, theologians have been mostly silent on this question, "no doubt feeling that the problems of earth are more than enough to occupy them."

No true astrotheology was developed in the 20th century in the sense that new theological principles were created, or existing ones formally modified, to embrace other moral agents in the universe. Although the mere possibility of extraterrestrial intelligence generated sporadic attempts at a universal theology, systematic astrotheology will probably be developed only when – and if – intelligence is discovered beyond the Earth. In the meantime, merely posing the problem demonstrates the anthropocentricity of our current conceptions of religion and theology and suggests that they should be expanded beyond their parochial terrestrial bounds. Though theologians have gone some way toward addressing Clarke's challenge, the theological legacy of the Space Age in a broader sense is still unfulfilled. And as C. S. Lewis suggested, if extraterrestrials are actually discovered, the problem will become much more urgent.

In the end, the effect on theology and religion may be quite different from any impact on the narrow religious doctrines that have been discussed during the 20th century. It may be that in learning of alien religions, of alien ways of relating to superior beings, the scope of terrestrial religion will be greatly expanded in ways that we cannot foresee. It may even be that, as a search for superior beings, the quest for extraterrestrial intelligence is itself a kind of religion. Even Puccetti argued that if extraterrestrials doomed religions, the religious attitude – the striving for "otherness," for something beyond the individual that offers understanding and love – might survive. "It is in that direction, and not in the direction of sophisticated machines or theological abstractions, that a plausible 'otherness' lies. Thus the religious attitude, or an important element of it, may yet survive the death of all determinate religious beliefs." It may be that religion in a universal sense is defined as the never-ending search of each civilization for others more superior than itself.

If this is true, then SETI may be science in search of religion, and astrotheology may be the ultimate reconciliation of science with religion.

8.3 LIFE AND PURPOSE IN THE UNIVERSE: THE ANTHROPIC PRINCIPLE

In the second half of the 20th century, as the search for extraterrestrial life reached new heights, another approach to the meaning of life came from an unexpected quarter: cosmology itself. This approach, known as the "anthropic principle," was first stated in 1961 (though not under that name), was elaborated during the 1970s, and by the last decade of the century was a topic of considerable discussion among cosmologists and philosophers. In some ways a counterpoint to the "theistic principle" that God designed the universe, the anthropic principle found significance in the fact that the universe in its deepest structure and most fundamental properties seemed tailor-made for humanity; in order to have produced life and intelligence, those fundamental properties could not have been much different from those we now observe. A link between the physical and biological universes, between mind and matter, the anthropic principle may be seen as a secular search for the meaning of life based on physical principles rather than theological dogma. Transcending the usual critical but more parochial concerns about local conditions for life, it gave the extraterrestrial life debate a cosmological component of the broadest possible scope.

Though the suitability of the cosmos for life is clearly an argument that applies to life anywhere in the universe, it is revealing that in its early formulations, at least, the idea was applied only to humanity (thus the term "anthropic"). Contemplating the work of Arthur S. Eddington, Paul Dirac, and others on certain coincidences in the relations between fundamental physical constants, in 1961 the Princeton physicist Robert Dicke pointed out that life could not exist at any random time in the history of the universe, but only after the universe had reached a certain age. Thus, the physical constant known as the "Hubble age" of the universe "is not permitted to take on an enormous range of values, but is somewhat limited by the biological requirements to be met during the epoch of man." Among those requirements was that the universe must be old enough for carbon to have been formed inside stars and then distributed, since physicists and their fellow humans are carbon-based. And the universe must be young enough to provide a hospitable home in the form of a planet circling a stable star, yet not so old that the inevitable stellar death precludes life. Thus, Dicke concluded (arguing against Dirac's idea that the fundamental "constants" might vary with time), the age of the universe as observed by us is not random, "but is limited by the criteria for the existence of physicists."

This early formulation of the idea that our very existence tells us something about the fundamental constants, though presented only 1 year after Drake's Project Ozma had raised consciousness about extraterrestrial intelligence, still spoke only in terms of how the "epoch of man" constrained the age of the universe. Dicke need not in fact have generalized, since nothing was known about the existence of extraterrestrials, and the argument derived its force from the existence of the one set of intelligent observers known with certainty to exist. But Dirac's reply in the same issue of *Nature* indicated that he did not agree with Dicke's argument precisely because it placed limits on life in the universe. On Dicke's assumption of nonvarying constants, he wrote, "habitable planets could exist only for a limited period of time. With my assumption they could exist indefinitely in the future and life need never end. There is no decisive argument for deciding between these assumptions. I prefer the one that allows the possibility of endless life."

Despite Dirac's objection, however, Dicke's concept was revived more than a decade later, and in such anthropocentric terms that it was now named the "anthropic principle." The occasion was a 1973 symposium on "Confrontation of Cosmological Theory and Observational Data," in which the Cambridge University physicist Brandon Carter expanded on Dicke's idea. Influenced by his reading of Hermann Bondi's *Cosmology* (1959), Carter pointed out that the anthropic principle could have been used to predict certain "large number coincidences" in cosmology that Bondi described, following in the footsteps of Eddington and Dirac. In its simplest form, dubbed the "Weak Anthropic Principle," the idea was that "what we can expect to observe must be restricted by the conditions necessary for our presence as observers." Thus, Carter cited Dicke's example that the age of the universe is constrained by the fact that we exist. Moreover, he went on to expound a Strong Anthropic Principle based on other constants, namely, that "the Universe (and hence the fundamental parameters on which it depends) must be such as to admit the creation of observers within it at some stage," a much more problematic claim. And he pointed out that the gravitational constant G was also critical to the existence of life, for if it were slightly weaker only red stars would form, whereas if it were slightly stronger only hot blue stars would form, neither being able to harbor life-supporting planets. In an ensemble of universes with all conceivable combinations of fundamental constants, Carter wrote, the existence of an observer will be possible only for certain restricted combinations of parameters. Although such reasoning turned the deductive method on its head, Carter argued that it might be considered a kind of explanation for why our universe is the way it is.

Without addressing the question of extraterrestrial life, Carter viewed the anthropic principle as "a reaction against exaggerated subservience to the 'Copernican Principle'." Carter also cautioned that just because we do not occupy a privileged *central* position in the universe, this does not mean that our position cannot be privileged in any sense. Even if there are other life-forms in the universe, our position in the universe as observers is a privileged one – at least as privileged as theirs. Thus, even if there were extraterrestrials (a point that Carter did not discuss), the anthropic principle was conceived from the beginning as a step away from Copernicanism and toward anthropocentrism.

Going one step further, and drawing on his pioneering role in quantum physics, John Wheeler proposed a Participatory Anthropic Principle, in which the observer has an intimate relationship with the origin of the universe. "Has the universe had to adapt itself from its earliest days to the future requirements for life and mind?" Wheeler asked, accustomed from quantum physics to statements that made no common sense. "Until we understand which way truth lies in this domain, we can very well agree that we do not know the first thing about the universe." It is possible, Wheeler suggested, that in the same way that quantum physics found the observer and the observed to have a close and totally unexpected relationship, the study of the physical world "will lead back in some now-hidden way to man himself, to conscious mind, tied unexpectedly through the very acts of observations and participation to partnership in the foundation of the universe." In its preoccupation with the observer and the observed, Wheeler viewed the anthropic principle as analogous to something like the uncertainty principle in quantum mechanics.

While some began to wonder whether the anthropic principle was really saying anything at all, cosmologists especially could not resist developing the idea further. By 1979 B. J. Carr and and Martin Rees showed how the anthropic principle could be used to determine most of the fundamental constants of physics; in other words, life was remarkably sensitive to cosmologically related numerical values. But at the same time, they realized that from a conventional physical point of view, the anthropic explanation of these coincidences was unsatisfactory, in part because it was based on an unduly anthropocentric concept of the observer. "The arguments invoked here assume that life requires elements heavier than hydrogen and helium, water, galaxies, and special types of stars and planets. It is conceivable that some form of intelligence could exist without all of these features – thermodynamic disequilibrium is perhaps the only prerequisite that we can demand with real conviction." Still, in these terms, the existence of humanity outweighed the uncertainty about exotic extraterrestrials, thus focusing on humanity as the only known center of the biological universe.

The idea that life was very sensitive to the values of the fundamental physical constants need not have been anthropocentric in the sense that it implied, particularly in its stronger forms, that life was somehow essential to the cosmos. To the extent that the only life we know of *is* terrestrial, and that we consequently know nothing about the properties of any potential extraterrestrial life, the focus on humanity was perhaps justified. But anthropocentrism could be carried to greater extremes in this context, and was in the hands of physicists J. D. Barrow and Frank Tipler, whose massive volume *The Anthropic Cosmological Principle* (1986) provided by far the most detailed treatment of the subject. In this volume, the concept became strongly anthropocentric in conjunction with arguments that extraterrestrial life did not exist. Placing the idea of an anthropic principle in the context of a long history of design arguments in theology, Barrow and Tipler argued – contrary to the opinion of the vast majority of contemporary scientists – that teleology (the search for purpose) may have a role in modern science, especially in biology. Having laid the foundations for their anthropic principle, they detailed that teleological role in physics and astrophysics, cosmology, quantum mechanics, and biochemistry. And they pointed out that the principle they were championing was a much more general version of the "fitness of the environment" type of argument proposed by the Harvard biochemist Lawrence J. Henderson seven decades before.

This general point of view would seem to favor life in the universe, since it meant that life could arise anywhere that both the fitness of the environment and the "fitness of the universe" were favorable. But in fact, the anthropic principle had more ambiguous implications, and in the hands of Barrow and Tipler was used to argue against extraterrestrial life. Carter had first emphasized that the time for the evolution of intelligence on the Earth (5 billion years) was within a factor of 2 of the entire time available before life around a main sequence star would become impossible (10 billion years). In other words, if the average time for the evolution of intelligence in the universe exceeded by only a factor of 2 that evolutionary time on Earth, intelligence could not evolve elsewhere in the universe. Barrow and Tipler accepted this as an argument against extraterrestrial intelligence, though they emphasized that a testable prediction of Carter's claim was the uniqueness of terrestrial intelligence in the Galaxy, and that his argument would collapse if any SETI program made a successful detection. They also pointed out that while SETI proponents believed that intelligence was an inevitable outcome of the evolutionary process, many evolutionary biologists did not agree with this application of teleology. Thus they held that intelligence on Earth was not foreordained and that it may not have occurred anywhere else in the Galaxy. The presence of intelligence on Earth, they argued, can be understood only by the weak anthropic principle, namely, that "only on that unique planet on which it occurs is it possible to wonder about the likelihood of intelligent life." In conjunction with the "space travel" ("Where Are They?") argument against extraterrestrial intelligence, the anthropic principle for these physicists became truly an anthropocentric view of the universe. Ruling out extraterrestrials left nothing but humanity as the measure of the universe. Indeed, anthropocentrism may be seen as the main argument of their volume as a whole, since the space travel argument was otherwise irrelevant to the anthropic principle. In the quarter-century from Dicke's first enunciation of the concept in 1961 to the publication of Barrow and Tipler's volume, the anthropic principle was transformed from an incidentally anthropocentric to a purely anthropocentric concept. That this occurred as proponents of extraterrestrial life were becoming more vociferous is perhaps no coincidence.

Although not everyone agreed with an extreme anthropocentric form of the anthropic principle, by the 1980s it was a well-established if still controversial subject that found a place in cosmology texts, as well as in popular literature. Commenting on the unusual post hoc mode of reasoning, the philosopher George Gale found the utility, and even the legitimacy, of the principle still questioned by both philosophers and cosmologists as of 1981. The philosopher Kenneth Winkler believed anthropic reasoning was based on concealed anthropocentric assumptions, and wondered further why we should even broach the idea that life could bring about an adjustment of physical constants if we had no idea how this could be done. Others rendered even harsher judgments. Heinz Pagels found it "a farfetched explanation for those features of the universe which physicists cannot yet explain. Physicists and cosmologists who appeal to anthropic reasoning seemed to me to be gratuitously abandoning the successful program of conventional physical science of understanding the quantitative properties of our universe on the basis of universal physical laws," he recalled. Pagels found the interminable debate about the status of the principle to be symptomatic of its chief weakness – there was no way to test it. He found it to be "needless clutter in the conceptual repertoire of science" and claimed that its influence had been sterile. Life is not a principle acting on the laws of nature, he wrote, but rather a consequence of them. Contrasting the anthropic principle with the theistic principle, Pagels noted that because many scientists were unwilling to accept God, the human-centered principle was "the closest some atheists can get to God."

Finally, the astronomer Fred Hoyle undermined the whole anthropic program when he concluded that, although the occurrence of life is the greatest of all problems, "It is not so much that the Universe must be consistent with us as that we must be consistent with the Universe. The anthropic principle has the position inverted." Instead of cosmology telling us about life, he suggested, we should let life (which we know more about) tell us about cosmology. Like the UFO debate and other controversies we have seen repeatedly in the history of the extraterrestrial life debate, the controversy over the anthropic principle is yet another case demonstrating the diversity of scientific cultures, particularly when it comes to the scientific method.

Despite the criticism, the anthropic principle continued to be discussed by both philosophers and cosmologists as the century entered its last dozen years. Reviewing Barrow and Tipler's book, the cosmologist G. F. R. Ellis questioned many of their claims but concluded that they had raised important and stimulating questions. Foremost among them was the importance of life and mind in the universe, which he now labeled as "of central concern in any complete understanding of cosmology." Carter himself elaborated his opinion by arguing that the anthropic principle was most powerful in the biological, and particularly the evolutionary, realm, where he used it to argue that intelligent life would be rare even on planets with favorable environments. Among the philosophers, Patrick Wilson argued that the principle was not anthropic at all. In its emphasis on humanity as the central focus of the universe, he found it more of a religious principle than a scientific one, "either a rather trivial truth or an impotent teleological principle." John Leslie, in a volume summing up two decades of interest in the philosophical status of the anthropic principle, concluded that the evidence that the universe was fine tuned was "impressively strong," that "the God hypothesis" (although perhaps not with its traditional meaning) was a strong one for explaining this fine-tuning, and that unless life is a fluke, "God is real and/or there exist vastly many, very varied universes."

Despite claims that it was tautological, illogical, or irrelevant, the anthropic principle showed no signs of disappearing from scientific or philosophical discourse by the end of the century. To the contrary, it was a major point of discussion in the Venice Conference on Cosmology and Philosophy, held in 1988, where at least one participant presented it as an argument in favor of extraterrestrial life: "If moderately uniform conditions exist and life can exist *somewhere* (at suitable times), it can almost certainly exist *almost everywhere;* and this implies infinite repetition of Life forms in an infinite universe." Inevitably, the anthropic principle was also used for theological purposes by those who saw no contradiction in the evidence of design in the universe; for them it was a more scientific and sophisticated version of the traditional "design argument" for the existence of God. In conjunction with Barrow and Tipler's use of the anthropic principle, at the end of the century one could therefore choose from the full spectrum of possibilities in the context of the extraterrestrial life debate: a positive argument, a negative argument, and the extraterrestrially neutral argument from design. But it is remarkable that just when anthropocentrism seemed irretrievably banished from the repertoire of reputable worldviews, it returned in a form more sophisticated than but remarkably similar to that of A. R. Wallace, who in arguing against the plurality of worlds at the beginning of the century concluded that "the supreme end and purpose of this vast universe was the production and development of the living soul in the perishable body of man."

Although the anthropic principle bears on the relationship between mind and the cosmos, unless one takes seriously Carter's argument that the time for the evolution of intelligence *may* on average be longer than the time scale when life will be possible (a very big "may"), it says nothing about the abundance of life in that cosmos. In this sense, it adds nothing new to the extraterrestrial life debate. But the anthropic principle does remind us that in the broadest possible sense, conditions are such in our universe that the fundamental constants give rise to stars, planets, and life, and that this state of affairs has not been, and may not always be, true for the entire history of the universe. Thus, even while keeping veiled the question of life's abundance, it gives added meaning to the importance of life and mind in the universe. At the same time, the anthropic principle tells us that there can be no extraterrestrial intelligence fundamentally different from humanity in our universe, in the sense of evolving outside the constraints imposed by the physical constants. This, indeed, still gives scope for a tremendous diversity of life and intelligence.

Finally, the elegantly and tellingly misnamed anthropic principle focuses on the question of purpose in the universe and thus on the issues of life and mind. Although most scientists shy away from questions of purpose, most are also willing to concede that we do not yet know everything, and that the science of the 21st or the 30th century may be vastly different in both method and content. It was the problem of purpose that the Nobel Prize-winning physicist Steven Weinberg broached when he wrote in a widely quoted pessimistic passage, "The more the universe seems comprehensible, the more it also seems pointless." That view did not take into account the possibilities inherent in the biological universe, where intelligence, whatever else its characteristics, is likely to be purposeful by definition. What that purpose may be we have not yet the slightest inkling, but if there is a meaning to life on Earth, it is undoubtedly linked ultimately to intelligence in the universe, if such exists. Whether or not life is found beyond the Earth, having posed the question of purpose in a cosmic context is itself a significant contribution of the extraterrestrial life debate in the 20th century.

9

SUMMARY AND CONCLUSION
The Biological Universe

The discussions in which we are engaged belong to the very boundary regions of science, to the frontier where knowledge, at least astronomical knowledge, ends, and ignorance begins.

William Whewell (1853)

The properties of matter and the course of cosmic evolution are now seen to be intimately related to the structure of the living being and to its activities; they become, therefore, far more important in biology than has been previously suspected. For the whole evolutionary process, both cosmic and organic, is one, and the biologist may now rightly regard the universe in its very essence as biocentric.

Lawrence J. Henderson (1913)

Unlike most of science, this topic extends beyond the test of a well-framed hypothesis; here we try to test an entire view of the world, incomplete and vulnerable in a thousand ways. That has a proud name in the history of thought as well; it is called exploration.

Philip Morrison (1985)

The universe is not the inert cosmos of the physicists, with a little life added for good measure. The universe is life, with the necessary infrastructure around; it consists foremost of trillions of biospheres generated and sustained by the rest of the universe.

Christian de Duve (1995)

When the 20th century began, a universe filled with life was widely accepted, completely unproved, and burdened with a checkered history. By the end of the century, scientists and the public embraced the biological universe even more enthusiastically, now in the context of an enormously larger universe. Still, nowhere among the innumerable stars and the fleeing galaxies had life been found. In the interim, as documented in this study, the best attempts of science to answer this age-old question yielded only circumstantial evidence in the solar system, tantalizing hints among the stars, and a revealing picture of the scientific enterprise and popular hopes and fears on Earth. Here we summarize the nature and significance of a debate that proponents and critics alike have viewed as crucial to humanity's self-image while remaining at the outer limits of the capabilities of science.

9.1 THE TRIUMPH OF COSMIC EVOLUTION

Despite the lack of a final answer about life in the universe, the 20th-century extraterrestrial life debate above all else demonstrates, as no other scientific controversy does, how completely the idea of cosmic evolution has triumphed, notwithstanding the continued controversy over terrestrial biological evolution led by those with vested religious interests. It is, indeed, this idea that unites all the seemingly disparate elements of the debate. Percival Lowell understood at the beginning of the century that the solar system had evolved and was evolving; his compelling picture of a dying Mars, whose inhabitants were desperately trying to distribute their water resources, was the epitome of a solar system constantly subject to change. The Space Age failure to detect vegetation or even organic molecules on Mars (at least before the Martian meteorite controversy), while a temporary setback to the cause of cosmic biological evolution, was, after all, a conclusion limited to a single planet. That innumerable planets might exist was an implication of the nebular hypothesis, whereby the birth of planetary systems was a natural by-product of star formation. Despite a temporary eclipse and Jeans's accompanying claims of planetary rarity during the 1920s and 1930s, by the 1940s the nebular hypothesis was revived, increasingly elaborated, and widely accepted throughout the remainder of the century, even as astronomers sought observational evidence of planetary systems, a search that brought confirmed successes only at the very end of the century. On such planets throughout the universe, it was postulated, chemical evolution and the origins and evolution of life had taken their own course, perhaps similar in a general way to those on Earth but subject, according to the principles of natural selection, to the environmental conditions on each planet. Beginning in 1953, Miller–Urey-type experiments demonstrated that the building blocks of life, if not life itself, could be produced under presumed primitive Earth conditions. By 1975 it was confirmed that nature itself had produced some of these building blocks in a variety of outer-space environments ranging from meteorites to interstellar molecules. Though by the end of the century the Oparin–Haldane model for the origin of life was under fire in details ranging from the nature of the first living organism to the constituents of the primitive Earth atmosphere, its essential premise that biological evolution on Earth followed chemical evolution as night follows day was not only widely accepted but had been extended to the universe at large.

The Viking missions failed in their local test to detect microbial life, but the goal of SETI was to test for the ultimate product of this hypothesis of cosmic evolution – extraterrestrial intelligence. The importance of the Drake Equation was not that it gave any definitive answer of the number of communicative technological civilizations in the Galaxy – a role even its users denied – nor that it was a tool around which discussion could focus. Rather, it was the very embodiment of the concept of cosmic evolution. Each of its disparate elements – individually uncertain and collectively almost devoid of meaning in any observational sense – represented one of a series of steps in astronomical, biological, and social evolution. Despite the uncertainties in astronomical and biological evolution, the fact that the number of technological civilizations hinged on L – the average lifetime of a technological civilization – emphasized that social evolution was the most crucial parameter of all. It was also the most unknown. The Drake Equation was not a puzzle to be solved for the century but a problem for the centuries to follow; its chief contribution for our age was to codify how cosmic evolution had become the paradigm of modern times.

Cosmic evolution became a paradigm not just in some vague and abstract way, but as a real research program, with NASA as its flagship patron and a community of researchers and enthusiasts around the world. It did so at first in piecemeal fashion, beginning with concern about planetary contamination and back-contamination, then more directly with the search for life on Mars, leading to broadly conceived research on the origin of life. When Viking failed to detect life on Mars, the cosmic evolution program expanded beyond the solar system to the search for planetary systems and finally to the search for radio signals as part of the SETI program. SETI, more than any other single activity, embodied all the elements of cosmic evolution in a unified research program explicitly enunciated by NASA, as graphically shown in Figure 9.1. And when SETI research was privatized, the NASA Origins program continued the program of cosmic evolution, encompassing not only the fields of exobiology and the somewhat more broadly defined astrobiology, but also everything from the Big Bang to modern humanity. The cosmic evolution program was viewed by many researchers around the world as the cosmic context of their work, whether in planetary science, planetary system science, or origin of life studies. Aside from sporadic interest, mainly in the Soviet Union (which by the 1990s itself dramatically demonstrated social evolution on Earth), only toward the end of the century was there a slowly awakening interest in social evolution in a cosmic context.

The concept of cosmic evolution was not a product exclusively of the Space Age, but it was the Space Age that brought it – and all its consequences for humanity's status in the universe – to the forefront as a research problem susceptible to science. Although the Harvard biochemist L. J. Henderson, for one, clearly grasped the idea (and even employed the terminology) of cosmic evolution at the beginning of the century, his sentiment, as stated in the passage at the beginning of this chapter, was not taken to heart in the first half of the century. Spurred by the Space Age, only during the last half was cosmic evolution embraced by an increasing number of scientists from many disciplines. But whether we may yet "regard the universe in its very essence as biocentric" remained the essence of the conundrum at the end of the century, whether in the scientific sense of the evolution of life, the philosophical sense of the anthropic principle, or the religious sense of the meaning and purpose

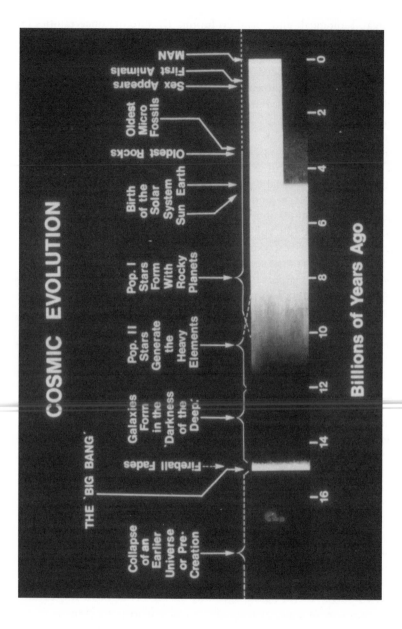

Fig. 9.1. Cosmic evolution, as depicted in this NASA drawing, encompasses events from the Big Bang to intelligence. NASA was the chief patron of research in cosmic evolution. Courtesy NASA.

of life. While the 20th century proposed, elaborated, and tested the concept of cosmic evolution, the question of biocentricity is likely to occupy the 21st century as one of its chief scientific and philosophical problems.

9.2 THE BIOLOGICAL UNIVERSE AS COSMOLOGICAL WORLDVIEW

Based on the concept of cosmic evolution, those championing the biological universe held that planetary systems were common, that whenever and wherever conditions were favorable life would originate and evolve, and that this evolution would culminate in intelligence, even if not necessarily one similar to that on Earth. SETI enthusiasts believed further that extraterrestrial intelligence would develop the same facility with radio technology as terrestrials, and that the lifetime of technological civilizations was sufficiently long that a large number of civilizations coexisted in the universe – large enough for Earthlings to mount significant efforts to listen for any radio messages from outer space. In short, supporters of the biological universe favored optimistic estimates for all the parameters in the Drake Equation. The biological universe was more than an idea, more than another theory or hypothesis; it was sufficiently comprehensive to qualify as a worldview. Since according to SETI advocates it was testable, it was a *scientific* worldview perhaps best described as a cosmology. And because it combined the biological universe with the characteristics of the physical universe, it may be most accurately termed the "biophysical cosmology."

That the idea of extraterrestrial life and all its accompanying assumptions should be viewed in cosmological terms is not a framework that we need impose on history; several participants in the debate implicitly or explicitly recognized it themselves. Among the first was the former Harvard Observatory Director Harlow Shapley, who in 1958 described extraterrestrial life as the "Fourth Adjustment" in humanity's view of itself in the universe since ancient Greece. The astronomer Otto Struve – crucial to our story for his advocacy of abundant planetary systems and for his support of Project Ozma – agreed, viewing the question as a revolution comparable to the Copernican cosmology in the 16th century and the "galactocentric" cosmology in the early 20th century. Others gave it a similar status, including Project Cyclops author Bernard Oliver, who spoke of "Biocosmology" in his influential study of 1971.

Like other cosmologies, the biophysical cosmology made claims about the large-scale nature of the universe, most comprehensively that life was one of its basic properties. Adherents believed that life was not merely an accidental or incidental property but an essential one; whether this meant that it might eventually change our perception of the physical universe via some anthropic or (in Henderson's words) "biocentric" principle, in the way that Einstein's theory of relativity changed our concept of the physical universe, was still un-certain. Like all cosmologies, the biophysical cosmology redefined the place of humanity in the universe, now precluding the last hope for anthropocen-trism unless one took the strict (and not very plausible) interpretation of the anthropic principle that the universe was made entirely for humanity. And like other cosmologies, its advocates believed the existence of widespread in-telligence was testable, resulting in SETI programs supported by groups rang-ing from grass-roots space advocates (via the Planetary Society) to universities and government.

Viewed in these terms, the study of the extraterrestrial life debate would un-doubtedly benefit from an historical comparison with those other great shifts in the cosmological worldview of the past. Valid parallels might be drawn with other cosmologies with respect to the development and evolution of the idea, the arguments and strategies of its critics, and its widespread acceptance in the face of inconclusive evidence. Moreover, such a comparison might pro-vide insights into the implications of the biophysical cosmology, for one of the hallmarks of cosmologies is that they affect the broader culture, includ-ing philosophy, literature, and religion. Although many aspects of culture have already been affected to some extent by contemplating the mere possi-bility of extraterrestrial intelligence, they are likely to be affected even more if and when the biophysical cosmology is confirmed. Nor need the parallels be confined to physical worldviews; the Darwinian worldview, with its con-sequences for humanity's status and broad social implications, might serve as an even closer parallel. Indeed, the biophysical cosmology raises Darwinism to a cosmic context in which humanity is part of a much larger chain of being rather than simply at the apex of the apes.

9.3 THE PROBLEM OF EVIDENCE AND THE LIMITS OF SCIENCE

In attempting to understand why the question of extraterrestrial life has not been resolved in the 20th century, the factor that looms largest is the prob-lematic nature of evidence and inference. This is true whether we are deal-ing with the nearest planets of the solar system, the search for more distant planetary systems, or theories and experiments about the origin and evolu-tion of life relevant to both. It is true whether using ground-based telescopes, or spacecraft that actually land and undertake in situ observations, or even meteorites that have landed on Earth and been subjected to the most sophis-ticated laboratory analytical techniques modern science has to offer. And it applies equally to the visual, photographic, and spectroscopic techniques of astronomy, as well as to the variety of techniques employed in studies of the origin and evolution of life.

The most notorious example of the difficulties of evidence and inference is Lowell's claim that he had detected canals on Mars. But these difficulties neither began nor ended with Lowell, nor with the controversy he fueled, nor were they endemic only to the extraterrestrial life debate. The apparent and much applauded triumph of Antoniadi, who believed he had resolved the canals into dark patches by 1909, was in the end no triumph at all, for we now know that most of the claimed canals correspond to no dark features on Mars, so that even Antoniadi had overturned Lowell's claims only by accident. The so-called wave of darkening followed by sophisticated spectroscopic observations led many to believe that Mars at least harbored life in the form of vegetation. Even with the Viking landers, the evidence was ambiguous at first, and although a consensus was reached relatively quickly that the results were not biological in origin, a few still dispute the result; even more would like to return to Mars to search for past or present life in other locations. The search for planetary systems chronicles similar difficulties with observational evidence – now complicated by increased interaction with theory.

In the case of the origin and evolution of life, the difficulties were of a different nature. At stake was not so much the difficulty of observations as the inability to make them, at least directly, resulting in heavy reliance on inferences drawn from experiments. No one could observe the primitive Earth's atmosphere or the origin of life. Experiments had to be performed under conditions of a presumed primitive atmosphere, and even if one accepted that atmosphere, the products of prebiotic experiments were only suggestive, not definitive, in attempting to explain how life might have originated on Earth. Experiments were no more definitive in determining whether the origin of life was a common process, and what exactly this might tell us about the abundance of life on other planets. The pioneering and startling observations of complex organic molecules in space, and even the triumph of unambiguously demonstrating the presence of amino acids in carbonaceous meteorites, were, in the end, still only suggestive. Neither experiments on Earth nor nature's experiments in interplanetary and interstellar space produced the equivalent to life, only the prebiotic building blocks of life. In the end, all arguments about the origin of life on Earth and its abundance in the universe remained presumptions.

In both the specific arguments represented by each element in the Drake Equation and the Grand Argument represented by all the elements in the Drake Equation taken together, evidence and inference were the culprits preventing definitive solution. Beyond the solar system, with the exception of artificial radio signals that were the holy grail of SETI programs, all observations became indirect.

Finally, evidence and inference were inevitably grounded in certain methodological assumptions, and scientists, philosophers, and historians have pointed out that at least some of the assumptions made in the pursuit of exobiology and SETI may well be false. This is undoubtedly true, yet at the same time, advocates emphasized that the assumptions could not be proved false. In short, they were the assumptions of those who adopted the biophysical cosmology as their worldview and sought to test it by whatever means possible. While exobiological assumptions might be more grandiose than those found in most science, they were not qualitatively different from it, and those who termed exobiology a science without a subject would seem either to have mistaken the nature of science or to have had an agenda of their own. No forefront science could be sure that its subject existed until the observations were made or the experiments undertaken. From gravitational wave astronomy to the magnetic monopole and the search for the gene as a unit of heredity, over whatever period one chooses in the history of science, substantial progress was most often made by making bold assumptions and following leads to their sometimes dead-end conclusions. Objections to the contrary are interesting exemplars of the rhetoric of science, no more evident than in the case of the evolutionist George Gaylord Simpson, who coined the phrase "science without a subject" in connection with exobiology while arguing that money would be better spent on down-to-Earth research – such as his own.

The lack of a definitive answer to the question of life in the universe, due in part to the problematic nature of evidence and scientific inference, leads us to the conclusion that science has limits in its ability to resolve certain questions. The search for extraterrestrial life has remained at the very limits of science during the 20th century in the sense that the required techniques have been barely sufficient, leading to problematic evidence and stretching the normal principles of scientific inference. In this sense, then, our history may be seen as a case study of how science functions at its limits. Among the limited number of scientists who entered the boundary region, most preferred observation over theory, but they did not hesitate to use theory in the absence of observation to further their argument. James Jeans and planetary systems is a prime example. Where evidence was available, scientists took a variety of approaches to the same data, whether dealing with Mars, or planetary systems, or the implications of complex molecules in space. The distinction between optimists and pessimists in the search for life was sharply drawn. How scientists assumed their roles as optimists and pessimists is a subject for further study, but once they took these roles, the debate between those functioning at the limits of science and their colleagues doing more normal science was often vigorous, encompassing not only technical arguments but also broad issues of methodology and philosophy.

Despite all the problems of working in science at its limits, the charge of critics that exobiology had been marked by lack of progress is not sustainable. Compared to the beginning of the century, we now know that no intelligence or vegetation exists on either Mars or Venus; that not even organic molecules are to be found on Mars; that planets exist around a variety of nearby stars and that material exists around stars that may (or may not) form into planets; that interstellar space contains complex organic molecules that could serve as the basic building blocks for life; and that civilizations do not seem to be beaming messages toward us at the 21-cm frequency. There is still no definitive answer, and although the biological universe is not verifiable globally, it is verifiable light year by light year as observations advance outward in an expanding sphere of search space for particular frequencies and other sets of assumptions. In the sense that it may need to search the entire universe, SETI is perhaps the largest research program ever proposed. Although the answer is bound by the limits of scientific reasoning and technique, SETI is limited only by the size of the universe.

Finally, we must admit that our science may well be limited compared to that of other civilizations. If comparison is possible, one wonders whether the long-sought "objective knowledge" might be found at last by gleaning the common elements remaining after processing by many sensory and mind systems independently evolved throughout the universe. If so, many of the problems of interpretation and inference might be resolved or at least reach a new level.

9.4 THE CULTURES OF SCIENCE

The different approaches to the problem of life in the universe, the diverse interpretations of data, and the vigorous nature of the debate between protagonists and antagonists in the debate lead us to a conclusion about the nature of scientists, as distinct from the science they produce. In his classic work *The Two Cultures* (1959), C. P. Snow contrasted the sciences and humanities as two cultures, "literary intellectuals at one pole – at the other scientists . . . between the two a gulf of mutual incomprehension." Our study of the extraterrestrial life debate demonstrates more clearly than most other areas of science what is evident to anyone with a broad overview of scientific ideas and practice but often remains unspoken, especially by scientists themselves: that whether in terms of problem choice, of method, of inference, or whatever other characteristic one wishes to choose, science itself has many cultures. Again and again we have seen this throughout our study: in Lowell's interpretation of the canals as compared to others (Table 2.1); in the Martian vegetation controversy; in claims of planetary systems; in the unwillingness of some to take the UFO phenomenon seriously and in their diverse interpretations of the same data (Table 5.2); in the meaning of prebiotic experiments for the origin of life; and in the diverse estimates for success in SETI (Table 7.1).

269

These cultures of science undoubtedly operate at many levels, from the routine to the frontier. They may have been glimpsed by those who have studied "styles" of science, whether national or regional. But the idea of "culture" is at once more personal and runs deeper than "style"; it may involve not only methodology, but also worldview; not only the problems that scientists actually take up, but also the ones they are willing to consider taking up; and not only day-to-day research, but what are viewed as the ultimate goals of science. Just as Snow's concept served a purpose for its time, the time may be ripe for a study of the many cultures of science. In particular, the field of science studies, which has often emphasized the multiplicity of the material, social, and conceptual elements of "scientific culture," could use the extraterrestrial life debate to study more precisely the many cultures that compose science. A study of how these cultures internal to science originate, dynamically interact, and affect the production of scientific knowledge promises to shed new light on the nature of the scientific enterprise.

9.5 EXOBIOLOGY AS PROTOSCIENCE

Those who did take up the quest for extraterrestrial life functioned in a world often seen by their peers as of dubious scientific status and, at least in its early stages, perhaps not as science at all. Real people risked real careers to study a problem that critics ridiculed as without content. The question of whether or not they succeeded in forming a discipline may be academic (unless one is among those participants searching for funding), but what is clear is that exobiology was at least an emerging, if not yet emergent, science. To put it another way, exobiology (taken in its broadest sense to include SETI and the entire program of cosmic evolution, sometimes also referred to as "bioastronomy") was a protoscience in the most literal sense of the word. Using an astronomical metaphor, like protostars and protoplanets, exobiology was a system (in this case of concepts, techniques, and researchers) still in some state of chaos. No one could quite be sure whether, like stars and planets, the new discipline would coalesce, or whether, after a promising start, centripetal forces would cause it to dissipate before critical mass was reached.

The forces of dissipation for protoscience may be great, having to do chiefly with funding, and therefore with politics, and ultimately with a society's self-image. The demise of government funding for the NASA SETI program demonstrated the fragility of the protoscience of exobiology, showing at once how the irrationality of the political moment could affect any program, no matter how carefully planned, even as society itself debated the larger issue of practical research versus curiosity-driven research. If the adherents of practicality won out, the entire program of exobiology (and indeed much else in science) would be seen as irrelevant, and exobiology, along with other aspiring disciplines, would dissipate into the void. How much would be lost in the process was anyone's guess. Proponents of curiosity-driven research would point to the mold and penicillin, among many other examples, waxing eloquent about curiosity as the essence of humanity, while critics would emphasize the dire needs of the world. In this sense, the fate of exobiology was part and parcel of a much larger debate at the end of the century, whose outcome was unpredictable and indeterminate. What seems clear is that, unlike established sciences like physics, biology, and chemistry and many of their subdisciplines, exobiology had little chance of enunciating a rationale of practical application. Nothing, advocates would say, except for the knowledge and wisdom of the universe, long a theme in alien literature (Fig. 9.2) and a hope of SETI practitioners.

9.6 CULTURAL SIGNIFICANCE OF THE DEBATE

Illuminating as the extraterrestrial life debate is for understanding the triumph of cosmic evolution and the nature of science and its many cultures, it is clear from our study that its significance goes well beyond the relatively parochial boundaries of science. Although the idea of extraterrestrials clearly fascinated a wide audience as early as Bernard le Bovier de Fontenelle's phenomenal best-seller *Entretriens sur la pluralité des mondes* (Conversations on the Plurality of Worlds) in 1686, no one could have foreseen the extent to which the idea would pervade popular culture by the 20th century. H. G. Wells's *War of the Worlds* (1897) and Kurd Lasswitz's *Auf zwei Planeten* (On Two Planets, 1897) were only the vanguard of an enormous number of diverse treatments of the alien theme in science fiction. And in the real world, one might have predicted that the first wave of reports of UFOs in 1897 would be a relatively minor flap and no concern of the sophisticated 20th century. But the revival of the UFO phenomenon in 1947, its peak in the 1960s with attention from government agencies, and the continued bizarre reports of alien abductions as the century ended showed that the interest in aliens was far from limited to fiction. That interest was confirmed when films such as *ET*, *Close Encounters of the Third Kind,* and *Independence Day* became among the most popular in cinematic history. First in science fiction literature, then in the UFO debate, and finally through the emotionally powerful medium of film, the idea of extraterrestrials insistently appealed to the popular mind and became an integral part of popular culture.

More than this, the idea of extraterrestrials constituted for many people a worldview no less influential and appealing than the geocentric spheres and choirs of angels of the medieval worldview. The move from the physical world to the biological universe, ending with what we have called in scientific terms the biophysical cosmology, was one of the most distinctive traits of the 20th

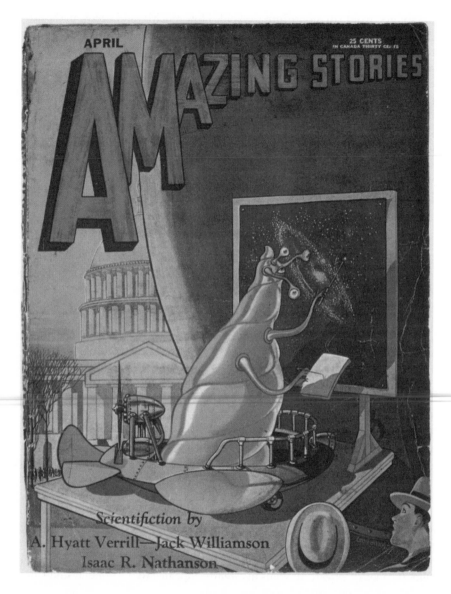

Fig. 9.2. Cover by Leo Morey for the April 1930 issue of *Amazing Stories*, showing an alien imparting knowledge of the galaxy. Copyright 1930 by Experimenter Publishing Co.

century, constituting what Harvard Professor Karl Guthke has called not only *a* myth, but perhaps *the* myth, of modern times. Although we normally associate mythology with something untrue, as Joseph Campbell and others have shown, mythology is in fact a reflection of the deepest beliefs and characteristics of a culture. Nor does this at all contradict the assertion that the biological universe constitutes a cosmology, a broad view of the world that itself may or may not be proved true. Whether the particular myth of our study is true we do not yet know, but that it has been claimed through different approaches to have the status of both myth and cosmology is some measure of its impact on modern culture.

Finally, the extraterrestrial life debate causes us to reexamine the very foundations of our knowledge and belief in a wide variety of subjects. Whether one chooses mathematics, religion, philosophy, science, the arts, or humanities, viewed from an extraterrestrial perspective they are all only local examples of what may perhaps be much more generalized knowledge and belief. All take on a new meaning if life is abundant in the universe – and if it is not. Even if extraterrestrials are not discovered, an extraterrestrially motivated reexamination of the knowledge and belief that we all too often take for granted is a considerable legacy for one of the 20th century's most persistent debates.

SELECT BIBLIOGRAPHICAL ESSAY

For complete references to all but the most recent events in this story, the reader should consult *The Biological Universe* (Cambridge and New York, 1996), of which this volume is the abridgment and update.

Among general works on extraterrestrial life, a succession of books has presented the knowledge of each generation in almost textbook-like fashion, beginning with Sir Harold Spencer Jones, *Life on Other Worlds* (New York, 1940), followed by I. S. Shklovskii and Carl Sagan, *Intelligent Life in the Universe* (San Francisco, 1966), and Donald Goldsmith and Tobias Owen, *The Search for Life in the Universe* (Menlo Park, 1980; 2nd edition, 1992). Primary readings related chiefly to the origin of life and life on Mars are reprinted in Elie Shneour and Eric A. Ottesen, *Extraterrestrial Life: An Anthology and Bibliography* (Washington, D.C., 1966), and the very early papers on what later became known as the Search for Extraterrestrial Intelligence (SETI) are reprinted in A. G. W. Cameron, *Interstellar Communication* (New York, 1963). Reprinted articles covering all these topics through the 1970s are found in Donald Goldsmith, *The Quest for Extraterrestrial Life: A Book of Readings* (Mill Valley, Calif., 1980).

Because no journal exists covering the broad range of extraterrestrial life studies, the largest single concentration of literature will be found in conference proceedings, as listed in Table 7.4. Eugene Mallove, Mary M. Connors, Robert L. Forward, and Zbigniew Paprotny, *A Bibliography on the Search for Extraterrestrial Intelligence* (NASA Reference Publication 1021, March 1978), with its categorization by publication, shows the extent to which the remaining literature is dispersed through an enormous number of popular and technical magazines and journals.

On the history of the debate, the Cambridge University Press trilogy began with Steven J. Dick, *Plurality of Worlds: The Origins of the Extraterrestrial Life Debate from Democritus to Kant* (1982), followed by Michael Crowe, *The Extraterrestrial Life Debate, 1750–1900: The Idea of a Plurality of Worlds from Kant to Lowell* (1986) and concluding with Steven J. Dick, *The Biological Universe: The Twentieth-Century Extraterrestrial Life Debate and the Limits of Science* (1996). Karl S. Guthke, *The Last Frontier: Imagining Other Worlds from the Copernican Revolution to Modern Science Fiction* (Ithaca, N.Y., 1990), is also a major contribution to the history of the debate, especially from the literary point of view. Also very useful is the popular account by the distinguished *New York Times* writer Walter S. Sullivan, *We are Not Alone* (New York, 1964; 2nd edition, 1993). All of these sources list numerous further references.

How we came to our modern view of the universe, the setting in which our story takes place, is described historically in Robert Smith, *The Expanding Universe: Astronomy's Great Debate, 1900–31* (Cambridge, 1982), and Richard Berenzden, Richard Hart, and Daniel Seeley, *Man Discovers the Galaxies* (New York, 1976). More popular treatments are C. A. Whitney, *The Discovery of Our Galaxy* (New York, 1971), and Timothy Ferris, *The Red Limit* (New York, 1977; revised edition, New York, 1983).

The history of solar system astronomy, which serves as background to Chapter 2, is given in Ronald E. Doel, *Solar System Astronomy in America: Communities, Patronage, and Interdisciplinary Research, 1920–1960* (Cambridge, 1995), and J. N. Tatarewitz, *Space Technology and Planetary Astronomy* (Bloomington, 1990). General problems of planetary observation are discussed in William Sheehan, *Planets and Perception: Telescopic Views and Interpretations, 1609–1909* (Tucson, 1988). The controversy over the canals of Mars is best covered in William G. Hoyt, *Lowell and Mars* (Tucson, 1976), in chapter ten of Professor Crowe's *The Extraterrestrial Life Debate,* and in a variety of more specialized studies cited in those studies. More about both Lowell and his observatory will be learned in David Strauss's forthcoming biography of Lowell.

Antoniadi's *La planète Mars* (Paris, 1930) is the apex of the visual tradition of Martian studies. Gerard DeVaucouleurs, *Physics of the Planet Mars: An Introduction to Areophysics* (London, 1954), holds the same place in the astrophysical tradition of Martian studies that Antoniadi's 1930 volume and Flammarion's *La planète Mars et ses conditions d'habitabilite* (Paris, 1892, 1901) hold for the visual tradition, each discussing all relevant observations to the time of publication.

The official history of the Viking mission is Edward C. Ezell and Linda Neuman Ezell, *On Mars: Exploration of the Red Planet, 1958–1978* (Washington, D.C., 1984). The view of one of the principal investigators of the biology experiments is given in Norman Horowitz, *To Utopia and Back: The Search for Life in the Solar System* (New York, 1986). A dissenting view by another of the principal investigators is given in Barry E. Digregorio, Gilbert V. Levin, and Patricia Ann Straat, *Mars: The Living Planet* (Berkeley, Calif., 1997). The final word on the canals of Mars in light of modern research is Carl Sagan and P. Fox, "The Canals of Mars: An Assessment after Mariner 9," *Icarus,* 25 (1975), 602–612. The latest research on Mars is described in massive detail in H. H. Kieffer et al. (eds.), *Mars* (Tucson, Ariz., 1992), and in a more popular manner in William Sheehan, *The Planet Mars: A History of Observation and*

Discovery (Tucson, Ariz., 1996). The debate over microfossils and other possible indicators of life in Martian meteorite ALH 84001 is covered in Donald Goldsmith, *The Hunt for Life on Mars* (New York, 1997).

On the history of theories of the origin of the solar system, see especially Stephen G. Brush, *A History of Modern Planetary Physics* (Cambridge, 1996), and Stanley L. Jaki, *Planets and Planetarians: A History of Theories of the Origin of Planetary Systems* (Edinburgh, 1978). On the discoveries of extrasolar planets beginning in 1995, see Dennis L. Mammana and Donald W. McCarthy, Jr., *Other Suns, Other Worlds?* (New York, 1995), Donald Goldsmith, *Worlds Unnumbered: The Search for Extrasolar Planets* (Sausalito, Calif., 1997), and Ken Croswell, *Planet Quest: The Epic Discovery of Alien Solar Systems* (New York, 1997). NASA's plans are described in *A Road Map for the Exploration of Neighboring Planetary Systems (ExNPS)*, ed. C. A. Beichman (Washington, D.C., 1996).

On the history of science fiction, I have found indispensable the volume edited by John Clute and Peter Nicholls, *The Encyclopedia of Science Fiction* (New York, 1993). Another volume that I have used extensively is Neil Barron, ed., *Anatomy of Wonder: A Critical Guide to Science Fiction* (New York, 1987), a complete guide to science fiction and its historical and critical studies. Other useful sources on the history of science fiction include Brian W. Aldiss, *Billion Year Spree* (New York, 1973), and its revision by Brian W. Aldiss and David Wingrove, *Trillion Year Spree: The History of Science Fiction* (New York, 1986); James E. Gunn, *Alternate Worlds: The Illustrated History of Science Fiction* (New York, 1975); and Lester del Rey, *The World of Science Fiction: The History of a Subculture* (New York, 1979). Encyclopedic treatments include Brian Ash, ed., *The Visual Encyclopedia of Science Fiction* (New York, 1977).

Bibliographies of UFO literature include Lynn Catoe, *UFOs and Related Subjects: An Annotated Bibliography* (Washington, D.C., 1969), compiled by the Library of Congress for the U.S. Air Force in support of the Condon Study, and Richard Michael Rasmussen, *The UFO Literature: A Comprehensive Annotated Bibliography of Works in English* (Jefferson, N.C., and London, 1985). Histories of the UFO debate, largely limited to the United States, are found in David M. Jacobs, *The UFO Controversy in America* (Bloomington, Ind., 1975), and Curtis Peebles, *Watch the Skies! Chronicle of the Flying Saucer Myth* (Washington, D.C., and London, 1994). The U.S. Air Force investigations into the Roswell incident are *The Roswell Report: Fact vs. Fiction in the New Mexico Desert* (Washington, D.C., 1995) and *The Roswell Report: Case Closed* (U.S. Government Printing Office, Washington, D.C., 1997). The conference at MIT on alien abductions is documented in C. D. B. Bryan, *Close Encounters of the Fourth Kind: Alien Abduction, UFOs, and the Conference at M.I.T.* (New York, 1995). The reader may keep up with both sides of the UFO controversy in *The Skeptical Inquirer,* the *Journal of Scientific Exploration,* and the *Journal of UFO Studies.*

The two chief historical studies on the origin of life are John Farley, *The Spontaneous Generation Controversy from Descartes to Oparin* (Baltimore and London, 1977), and Harmke Kamminga, "Studies in the History of Ideas on the Origin of Life from 1860," Ph.D. thesis, University of London (November 1980). Robert Shapiro's *Origins: A Skeptic's Guide to the Creation of Life on Earth* (Toronto and New York, 1987) is a very readable guide to present theories on the origin of life. Current issues may be followed in the various conference proceedings and in the journal *Origins of the Life and Evolution of the Biosphere.* Also readable and thought-provoking is the volume by the Nobelist Christian de Duve, *Vital Dust: Life as a Cosmic Imperative* (New York, 1995). For the latest on self-organization in the context of the origin of life, see Stuart Kauffman, *At Home in the Universe: The Search for the Laws of Self-Organization and Complexity* (New York and Oxford, 1995). Many of the pioneering papers on the origin of life are reprinted in David W. Deamer and Gail R. Fleischaker, *Origins of Life: The Central Concepts* (Boston, 1994).

SETI is still a young enough science that its history has never before been written. Much source material will be found in David W. Swift, *SETI Pioneers: Scientists Talk about Their Search for Extraterrestrial Intelligence* (Tucson, 1990), 49–53. The history of NASA's involvement in SETI is elaborated in detail in Steven J. Dick, "The Search for Extraterrestrial Intelligence and the NASA High Resolution Microwave Survey (HRMS): Historical Perspectives," *Space Science Reviews,* 64 (1993), 93–139, on which Chapter 7 is based in part. A fascinating inside account from the point of view of one of the SETI pioneers is F. Drake and D. Sobel, *Is Anyone Out There? The Scientific Search for Extraterrestrial Intelligence* (New York, 1992). A witty and clear account of SETI is given in G. Seth Shostak, *Sharing the Universe* (Berkeley, Calif., 1998). A prime example of how a substantial interest in SETI has spread to other countries is Guillermo A. Lemarchand, *El Llamado de las Estrellas: Busqueda de Inteligencia Extraterrestre* (The Call of the Stars: Searching for Extraterrestrial Intelligence) (Buenos Aires, 1992). A psychologist critiques SETI's assumptions in John C. Baird, *The Inner Limits of Outer Space* (Hanover and London, 1987), an unusual contribution because its author is a social scientist.

Little scholarly literature exists dealing specifically with the implications of discovering extraterrestrial intelligence. The chief exceptions are Albert A. Harrison, *After Contact: The Human Response to Extraterrestrial Intelligence* (New York, 1997), and Paul Davies, *Are We Alone? Philosophical Implications of the Discovery of Extraterrestrial Life* (New York, 1995). Popular treatments have appeared in James Christian, ed., *Extraterrestrial*

Intelligence: The First Encounter (Buffalo, N.Y., 1976), and B. Bova and Byron Preiss, *First Contact: The Search for Extraterrestrial Intelligence* (New York, 1990). More serious discussions touching on the subject are found in Edward Regis, Jr., ed., *Extraterrestrials: Science and Alien Intelligence* (Cambridge, 1985), and in the NASA-commissioned study by John Billingham, Roger Heyns, David Milne, et al., *Social Implications of Detecting an Extraterrestrial Civilization: A Report of the Workshop on the Cultural Aspects of SETI* (SETI Institute Preprint, Mountain View, Calif., 1994). A collection of articles dealing mainly with the religious dimensions of the UFO phenomenon is James R. Lewis, ed., *The Gods Have Landed: New Religions from other Worlds* (Albany, N.Y., 1995).

The question of the limits of science is approached from several directions in Peter Medawar, *The Limits of Science* (Oxford, 1984); Nicholas Rescher, *The Limits of Science* (Berkeley, Calif., 1984); David Faust, *The Limits of Scientific Reasoning* (Minneapolis, 1984); Gerald Holton and Robert S. Morison, eds., *Limits of Scientific Inquiry* (New York, 1979); and Richard Morris, *The Edges of Science: Crossing the Boundary from Physics to Metaphysics* (New York, 1990).

The usefulness of historical analogues and the value of studying the reception of scientific worldviews are examined in I. Almar, "The Consequences of Discovery: Different Scenarios," and Steven J. Dick, "Consequences of Success in SETI: Lessons from the History of Science," both found in G. Seth Shostak, ed., *Progress in the Search for Extraterrestrial Life* (San Francisco, 1995), 499–505, 521–532.

INDEX

Abbot, Charles G.: and Martian vegetation, 44; and Venus, 4

Abelson, Philip H., 57

Academy of Sciences, USSR, 216

Adams, Walter S., 77f; on life on Mars, 44; oxygen absent on Mars, 48; water vapor absent on Mars, 48

Adamski, George, 148

Air Force, U.S., and UFOs, 138–139, 141–159

Aitken, R. G., and planetary systems, 78

alien: development of, 116–125; invention of, 107–116; see also film; life, extraterrestrial; science fiction; television; unidentified flying objects

ALH84001, see Mars, meteorites from

Allegheny Observatory, 91, 99

Alvarez, Luis, 147

Ambartsumian, V. A., 216

American Association for the Advancement of Science, 155–157, 158–159

American Astronomical Society, 83

American Institute of Aeronautics and Astronautics, and UFOs, 158

Ames Research Center, see National Aeronautics and Space Administration (NASA)

amino acids, 174; and comets, 183; and interstellar molecules, 183, 184; and life, 183, 191; in meteorites, 172, 180, 182–183, 184; in Miller–Urey experiment, 171, 184; synthesis of, 177, 191; see also organic synthesis, experiments in

Anders, Edward, 181

Antarctic, meteorites from, 1–2, 65–68

anthropic principle, 254–259; and Copernicanism, 255–256; and SETI, 257–258

anthropocentrism, decline of, 21–24; and Hubble Deep Field, 23f; and plurality of worlds, 21–24; and Russell, 82; and Shapley, 86; and Sun's position in space, 21–22; in Wallace, 21–22, 22f, 259; see also anthropic principle

Antoniadi, E. M., 29, 38–41, 38f, 267

Aquinas, Thomas, 9

Arecibo Observatory: message sent, 217; SETI observations, 223, 229–230, 225t

Aristotle, belief in a single world, 8–9

Army, U.S., 202f

Arnold, Kenneth, and UFOs, 140, 141

Arrhenius, Svante: and panspermia theory, 170, 179; and Venus, 43

Asimov, Isaac, 132

astrobiology, 68, 263; see also bioastronomy; exobiology

astrophysics, and extraterrestrials, 18

astrotheology, see theology

Atchley, Dana, 209

atmosphere, primitive Earth, 171, 173–174, 177, 191, 192; see also Mars

atomism, 7–8

bacteria, 66–67

Ball, Robert S., 21

Barnard's star, possible planets around, 82, 89–92, 91f, 93t

Barrow, J. D., 257–259

Beck, Lewis W., 240–241

Beckwith, Steven, 103

Belsky, T., 51

Bentley, Richard, 15

Berenzden, Richard, 238–240, 239f

Berkner, Lloyd: and Green Bank conference, 209; and Project Ozma, 205; and Space Science Board, 209; and UFOs, 147